SHE DOES MATH!

**Real-Life Problems
from Women on the Job**

CLASSROOM RESOURCE MATERIALS

Published by
THE MATHEMATICAL ASSOCIATION OF AMERICA

———

SHE DOES MATH!

Real-Life Problems
from Women on the Job

MARLA PARKER
Editor

Published and Distributed by
THE MATHEMATICAL ASSOCIATION OF AMERICA

All royalties from *She Does Math!* will benefit the Women and Mathematics Program of the Mathematical Association of America.

©1995 by
The Mathematical Association of America (Incorporated)
Library of Congress Catalog Card Number 95-76294

ISBN 0-88385-702-2

Printed in the United States of America

Current printing (last digit):
10 9 8 7 6 5 4 3 2

CLASSROOM RESOURCE MATERIALS

This series provides supplementary material for students and their teachers—laboratory exercises, projects, historical information, textbooks with unusual approaches for presenting mathematical ideas, career information, and much more.

Proofs Without Words, by Roger Nelsen

A Radical Approach to Real Analysis, by David Bressoud

She Does Math! Real-Life Problems from Women on the Job, edited by Marla Parker

Learn from the Masters! edited by Frank Swetz, John Fauvel, Otto Bekken, Bengt Johansson, and Victor Katz

101 Careers in Mathematics, edited by Andrew Sterrett, Jr.

MAA Service Center
P.O. Box 90973
Washington, DC 20036
1-800-331-1MAA   FAX: 1-301-206-9789

Preface

In today's competitive world, a good education is a necessity. By combining that education with a strong background in math and logic, you are ready for any career. I created this book for two reasons: to motivate students to take math every year in high school; and to encourage high school and college students—especially women and minorities—to consider technical fields when planning their careers.

Although women have always worked at home and in business, they are still a minority in most technical professions. For this reason, all the contributors to this book are women from many different fields. In their own words they tell how they became interested in math, where they went to school, and how they chose a career. The problems in each chapter are taken directly from their work experiences, to show how math and logic are used to solve real problems even in fields that aren't math-oriented, such as nursing. Here is a collection of concrete answers to the question, "Why should I take math?"

The women and men of past generations fought hard to win the right to be educated, to vote, and to work in any field they chose. Thanks to their efforts, I enjoy a challenging, well-paid job in a technical field. I hope this book will enable more women and minorities to enjoy these benefits also.

Marla Parker

Marla Parker and her daughter Jessica

About the Editor

Marla Parker was born in Houston, Texas. She received her BA in Computer Science from the University of California, Berkeley. She is currently a software engineering manager at SunSoft, A Sun Microsystems, Inc. Business. Her professional interests include code reuse and technologies that enable electronic commerce. She earned her private pilot license in 1991. She lives with her husband, daughter, a different foreign aupair each year, and two goldfish. She can be reached by email at: marla.parker@eng.sun.com.

Contents

Problems by Subject

No two individuals would associate the collection of problems found in this book with exactly the same subjects. However, these suggestions are offered with the knowledge that teachers will make their own assessments before any assignments are made.

Algebra   15–17, 21–24, 30–37, 38, 39–44, 45–47, 51–54, 58–60, 61–68, 69, 70, 72, 73, 82–84, 97, 103, 104–105, 106–111, 112, 113–119, 120–122, 125, 126–128, 136–140, 145–149

Astronomy   38, 144

Business   21–24, 30–37, 51–54, 70, 104–105

Calculus   69, 95, 97

Chemistry   106–111, 126–128

Computer science   11–14, 19, 48–50, 71, 74–77, 100–102, 123–125, 129–135

Geometry   15–16, 18–19, 27–29, 50, 78–81, 82–84, 145–149, 150–154

Health sciences   58–60, 61–68, 85–94, 113–119, 120–122

Home economics   85–94, 106–111

Piloting a plane   39–44, 78–81

Physics   25–26, 30–37, 38, 39–44, 45–47, 69, 72, 73, 120–122, 123–125, 141–143, 144

Puzzles   20, 49–50, 82–84, 103, 145–149

Statistics   1–10, 20, 55–57, 95–97, 98–99

Trigonometry   25–26, 69, 78-81, 112, 120–122, 141–143, 144

Susan C. Knasko

Environmental Psychology

Math, how boring! I was good at it, but didn't find it thrilling. That's how I felt in junior high school, when my career goals were to be a spy or an astronaut. (Already, I had several pamphlets from the FBI and had started the neighborhood Star Trek Spy Club.) My dad encouraged me to take a lot of math, reminding me that math is what got people to the moon.

In high school I studied geometry (I loved the logic), algebra, and trig-advanced algebra (I had *no idea* what those sines and cosines were for). Yike—before I knew it, I was a senior, and it was time to choose a college and a major. By now my goals had changed. I wanted to be a psychologist or a biologist. I didn't know what people in those professions really did, but my older sister Kathy had taken some psychology classes that sounded interesting, and my days in scouting and 4-H had taught me to love the outdoors.

College was fun. I could take classes in all sorts of fields. Some subjects that had been dull earlier turned out to be fascinating. For example, college-level geography involved flying around in airplanes, comparing ground sightings to aerial photographs and topographic maps—*not* memorizing state capitals. In a statistics course, the professor asked if I'd ever thought about becoming a math major. No, I still didn't find math exciting, although my roommate, who *was* a math major, seemed to enjoy it. I decided to major in psychology and minor in biology, sociology, and geography, because I loved them all and wanted a career that would combine them. When I told this to my advisor, he suggested a new branch of psychology called *Environmental Psychology*. He said it involved studying the relationship between the physical environment and human behavior. Yes, that was the field I would study in graduate school.

I didn't want to go straight to graduate school, so I took a few years off. First, I backpacked all over the country with a fellow psychology major. Then I worked for the U.S. Forest Service in the national forests of Pennsylvania, Nevada, and Idaho. What fun and what experiences! Finally, I went back to school and earned a master's and doctorate degrees in human-environment relations, taking many research design and statistics courses. My research focused on the effects of environmental stressors, especially noise, on human behavior and physiology.

After graduation, I came to the Monell Chemical Senses Center in Philadelphia, an institute that studies smell and taste. Here I investigate how ambient odors affect human health, mood, and behavior. Math plays a major role in my career, and I actually think the data analysis is the most fun part of an experiment. For months, I plan a project and collect data. Then comes the day when I can analyze the data to see whether my prediction was correct. In some ways, I did become the detective I always wanted to be. I became involved in the space program too, when I gave a talk about my research at one of NASA's research centers. What kind of career do *you* want? Just remember, it will probably require some math!

Environmental Stressors

Thirty people (referred to as subjects) participated individually in a study. Half were chosen because they were experts at a certain paper-and-pencil-task that requires concentration, and half were chosen because they had never performed the task before. When the subjects arrived, they were seated in a small room, given instructions for the task, and then left alone for an hour to work on it. During this time, an unpleasant odor was pumped into the room that held one-third of the experts and one-third of the beginners. A pleasant odor was pumped into the room holding another one-third of the experts and beginners. The remaining subjects in each group received only fresh air in their room.

At the end of the study the subjects completed a mood questionnaire. In addition, the number of errors made on the task were tallied for each subject, and are listed below.

Experts			Beginners		
Unpleasant Odor	Pleasant Odor	No Odor	Unpleasant Odor	Pleasant Odor	No Odor
8	7	5	32	27	14
9	2	6	25	16	17
0	4	1	19	31	13
3	1	3	29	11	15
6	2	5	8	26	13

The scores on the mood scale range from −24 to +24, where −24 = extremely bad mood, 0 = neutral mood, and +24 = extremely good mood. The mood scores for the subjects in each group are listed below.

Experts			Beginners		
Unpleasant Odor	Pleasant Odor	No Odor	Unpleasant Odor	Pleasant Odor	No Odor
6	12	6	7	10	5
−5	8	11	−10	16	−7
−2	0	2	21	5	9
14	17	−2	8	−4	−4
7	−1	6	−6	13	18

Problem 1. What is the average number of errors made by experts exposed to the unpleasant odor?

Problem 2. Which group made the largest number of errors?

Problem 3. Did either odor make beginners perform, on average, like experts?

Problem 4. Did exposure to an odor seem to have more effect on experts or beginners?

Problem 5. How might you summarize the performance (number of error) findings?

Problem 6. Were there strong differences in mood between beginners exposed to an unpleasant odor and experts exposed to the same odor?

Problem 7. Were there any differences between the odor conditions on the mood variable?

Problem 8. Considering the answer to Problem 6, would it be worthwhile to scent a room with pleasant odors for experts who needed to perform this task?

Problem 9. Considering both the error and the mood findings, which of the four scented groups might benefit most from having an odor in the room?

Problem 10. What further experiments might be conducted to explore the effects of odor on mood?

Mary E. Campione
Software Engineering; Computer Science

I grew up in a very traditional, military family. Girls became young ladies, married at an early age (before they were 20 years old), and had children. They did not learn mathematics or science, and certainly didn't go to college. Every time we got into a discussion about college, my parents would ask, "Why waste four years in college, and all that money, if you're just going to get married and have children?"

When I was about seven years old, these restrictions made me decide that being a girl was basically a bad thing but that fortunately, I was really a boy. So I dressed like a boy, played "army" and "fort" with boys, and did well in school—especially in science and math. To this day, I'm not sure how I managed to act like a boy for so long without my parents putting a stop to it. But they didn't. I finally relinquished my boyhood a few years later, but by then I was headstrong and adamant that I could do things girls weren't supposed to do.

All through high school I continued to earn straight A's, and participated in a number of sports. A teacher dropped a couple of hints that completely formed my choice of college and major. Once he said, "I think you'd be good at computer science." Weeks later he commented, "Cal Poly at San Luis Obispo has a good computer science program." So that's how I ended up majoring in Computer Science at San Luis Obispo.

During the first two years of college, my major courses were filled with about half women and half men. In the third and fourth years, the number of women in the math and computer science courses dwindled, until there were just one or two women in a class of 25 students.

In the summer after my sophomore year, I went to the San Francisco Bay Area to take a cooperative education position at a small company called Fortune Systems. For the first time, my family saw the value of college—after only two years of study, I was already making more money than my father!

I've been out of college for more than eight years now. The number of women in my field has grown, but even now I know of at least one company where the engineering organization is staffed with more than 80 men and only two women. I've worked for three different companies in Silicon Valley. My jobs have included programming, software design, technical writing, and supporting third-party software developers. While I was in the last of those "real" jobs, a partner and I started a software company that develops applications for the NEXTSTEP software platform. We work as contractors, using our technical writing, object-oriented design, and programming skills. We've also written two computer books.

Recently, I got married. Even though my parents are very proud of my accomplishments, I think they're relieved that I "finally" got married. I talked to my mother the other day—she still wants to know when we're going to have children!

Kerning

Kerning is a typographical term that means adjusting the space between letter pairs to make them both more readable and visually appealing. For example, the upper case letters 'A' and 'V' are often printed closer together than other letters, to take advantage of the similar slopes on their adjoining sides. For the most part, there are relatively few letter pairs that benefit from kerning in the English language.

For desktop-publishing computer systems that support kerning, the problem is how to store the kerning values (the space adjustment for each letter pair, measured in points) for these few special cases. Intuitively, kerning values would be stored in a 52 by 52 matrix (26 for lower case and 26 for upper case in each dimension). Like this:

```
        a    b    c   ...   z    A    B    C   ...   Z

a

b

c

:

z

A

B

C

:

Z
```

The cells in this matrix, known as a sparse matrix, would contain the kerning value for the letter pair that intersects at each cell. As we noted before, relatively few of the cells would have values, since kerning occurs infrequently. So the matrix would be filled mostly with zeros. Computer memory is expensive and is often a limited resource, so such waste is unacceptable. In fact, many theoreticians spend their careers developing ways to remove those last few bits of storage from data constructs.

Problem 11. Can you think of an alternative way to store only the important information from the kerning table? How much (in percentage) memory will you save? Can you think of any other ways? Which one is best? Why?

Palindromes

A palindrome is a word, phrase, or sentence that reads the same forward as it does backward. This sentence is a palindrome: "A man, a plan, a canal, Panama!" The word "redder" is also a palindrome.

Problem 12. Can you think of an algorithm to locate the center character (or two characters if an even number) of any given palindrome? If you know a computer language, use it—otherwise describe the algorithm in pseudo-code (English words formatted like a computer language).

The second part of the problem is: if there is an even number of characters in the palindrome, do the two characters located in the center always have to be the same?

Recursion and the Factorial Function

Computer languages contain objects known as subroutines—segments of code that are given a name, and then "called" from other segments of code. Recursion is what happens when a subroutine calls itself.

The factorial function in mathematics is noted in the following way:

$$n!$$

and means the product of all of the numbers from 1 to n, or:

$$n * (n - 1) * (n - 2) * \cdots * 2 * 1$$

Problem 13. Write a recursive subroutine that computes the value of $n!$. Then rewrite it without using recursion. If you don't know a computer language, use psuedo-code.

Recursion and Fibonacci Numbers

As I noted in the previous problem, computer languages contain objects known as subroutines—segments of code that are given a name, and then "called" from other segments of code. Recursion is what happens when a subroutine calls itself.

The following equations define the Fibonacci numbers (named after Leonardo Fibonacci, a 13th-century Italian mathematician).

$$\mathrm{fib}(1) = 1$$

$$\mathrm{fib}(2) = 1$$

$$\mathrm{fib}(n) = \mathrm{fib}(n - 1) + \mathrm{fib}(n - 2) \qquad \text{if } n > 2$$

You could use the equation for $\mathrm{fib}(n)$ to calculate the nth Fibonacci number, for example:

$$\mathrm{fib}(3) = \mathrm{fib}(2) + \mathrm{fib}(1) = 2$$

$$\mathrm{fib}(4) = \mathrm{fib}(3) + \mathrm{fib}(2) = 3$$

$$\mathrm{fib}(5) = \mathrm{fib}(4) + \mathrm{fib}(3) = 5$$

So the third Fibonacci number, $\mathrm{fib}(3)$, equals two. The fourth equals three, and the fifth equals five.

Problem 14. Write a recursive subroutine to compute Fibonacci numbers. Then use your subroutine, or the subroutine given in the answer, to calculate the seventh Fibonacci number, or any nth Fibonacci number you choose (but pick a small one unless you have lots of free time, or a computer to run the subroutine for you!)

Shelley J. Smith
Archaeology

Ever since I was in junior high school, I have been interested in archaeology and the mysteries of the past. It's fascinating to learn how archaeologists decipher the past from sparse clues—broken pottery, abandoned homes, trash middens—and how they employ many of the same strategies as detectives.

In school, I was neutral towards math; I didn't like it, but I didn't strongly dislike it either ... until high school, when I took geometry, trigonometry, and college algebra. I wasn't good at these subjects, and the C's and B's I received in these classes pulled my grade point average down. After that, I avoided taking any more math.

By tenth grade, I already wanted to be an archaeologist, so I took several elective humanities courses and read as much as I could about the subject. Archaeology is usually thought of as a social science, and in those days (the early 1970s), statistical analyses weren't widely utilized in the field. I felt very comfortable with my career choice, partly because it didn't necessitate taking higher math courses. During my undergraduate years, I was successful at avoiding higher math, earning a BA in anthropology from Pennsylvania State University in 1975.

But by the time I got to graduate school in 1981, there was no way around statistics and computer science courses. Archaeological fieldwork generates enormous amounts of data. Computer manipulation and statistical analyses of the data are the only means to make sense of it. Severe math anxiety was upon me! I had no choice but to nervously plunge in.

Then the most amazing thing happened: I enjoyed statistics and computer science. They were hard work, but my thesis research posed questions that only statistics could answer. For the thesis, I studied the artifacts from 476 Predynastic Egyptian tombs and 85 human figurines, to see if it was possible to learn about gender roles through these material remains. This study was fascinating, and computer programs enabled the management of large amounts of data that otherwise would have taken months to analyze. It felt good to tackle something I hadn't believed I could do. In 1984, I earned an MA in anthropology from Washington State University.

I could have made it much easier on myself if I had built on my high school math foundation with more math and science courses. Instead, I spent considerable time re-learning the math I had once known for my statistics and computer science courses.

Today, I am an archaeologist for the US Department of the Interior Bureau of Land Management, and have also worked for a consultant firm and university contracting divisions. As a Peace Corps volunteer, I worked in four western states and on the Caribbean island of St. Lucia.

Mapping a Dig

Archaeology is a destructive science—once a site has been dug it is gone forever, and can never be replaced exactly the way it was. The relationships of artifacts to each other (their "context") tell the story of prehistoric people, just as the collection of things in your room reveals much about your nature, interests, and the time and culture in which you live. However, if a certain item, say a picture you cherished, was removed from your room and placed somewhere else, a piece of your story would be missing. The picture would have been removed from its context, the place where it functioned.

Since archaeologists destroy a site when digging it, they preserve the context by recording it all on paper. Archaeologists can even study a site dug many years ago, if good notes and maps were made. Before archaeologists excavate (or "dig") a site where people once lived, they place a rectangular grid, called the Cartesian coordinate system, over the site surface. Each square on the ground is marked with numbered stakes in the corners, which give it a unique "name" that can be referred to by its coordinates, the place where the x axis and the y axis intersect. Any artifacts or samples found in that square are labeled with its grid number.

To lay out an accurate grid system, archaeologists use the Pythagorean theorem, $X^2 + Y^2 = Z^2$. The base lines, the x and y axes, are placed on the ground by using a survey instrument, such as a transit. (See Figure 1.) After deciding how big the quadrant should be, tape measures are attached to two ends of the baselines defining a rectangular quadrant. The origin, or $(0,0)$ point, is where the x and y axes intersect.

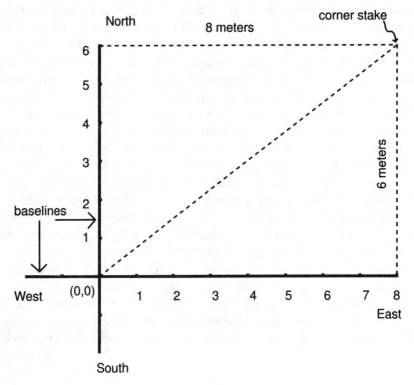

FIGURE 1
Grid system diagram for mapping a dig

The proper length of the diagonal is calculated using the Pythagorean theorem, to make sure the quadrant is a rectangle and not a lopsided parallelogram. A third tape measure is attached to the origin and stretched diagonally across the quadrant. The intersection of the three tape measures gives an accurate location for placement of a corner stake.

Problem 15. Calculate what the length should be for the origin-anchored tape measure. All measurements are metric.

Sizing Pottery from a Sherd

The most commonly-found type of artifact is the pottery sherd. Fired clay vessels are very durable and will last for thousands of years, even if they are lying on the ground surface. Since pottery styles are distinctive to particular groups of people, and the styles changed over time, pottery is a good way to determine how old a site is and what group of people lived there.

Archaeologists also want to know what certain pottery vessels might have been used for— were they for cooking, serving, or storing food? Since pots are usually found broken into hundreds of pieces, it is a tedious and often impossible job to glue them all back together.

To quickly get an idea of how large a pot was, calculate its circumference from a curved sherd. Using a rim sherd will tell how large the vessel opening was. Lay the rim sherd finished-edge down on a piece of paper, and trace the inside curve. Draw and measure a straight line between two points on the curve (this is a chord). At the midpoint of that line, measure the middle ordinate (the perpendicular distance to the curve).

C = chord length
M = middle ordinate

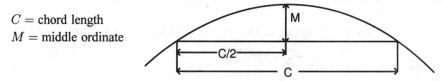

FIGURE 2
Sherd diagram for calculating the radius

The circumference formula is: $2\pi R$, where R is the radius. Unfortunately, in this case we don't know the radius, because there is only one section of the circle. But the radius of a circle or sphere can be determined by measuring a chord length (C) and the distance from the midpoint of the chord to the curve of a section of the circle (M). Redrawing the figure gives:

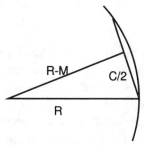

FIGURE 3
Determining the radius of a circle from a section

The Pythagorean theorem ($a^2 + b^2 = c^2$) can then be used to determine the radius, R, where

$$c = R \qquad b = R - M \qquad a = C/2$$

So:

$$R^2 = (C/2)^2 + (R - M)^2$$
$$R^2 = C^2/4 + R^2 - 2RM + M^2$$
$$R^2 - R^2 + 2RM = C^2/4 + M^2$$
$$2RM/2M = C^2/8M + M^2/2M$$
$$R = C^2/8M + M/2$$

Since the term $M/2$ has a small value, and since we are calculating only an approximate pot circumference, it can be neglected, giving the commonly used formula:

$$R = C^2/8M$$

Many clues about how a group of people lived can be extracted from data like this. If enough of a pot is present, its volume (or storage capacity) can be calculated. The storage capacity of vessels helps determine how much food people had stored, and from that, how many people lived at a site. Functions of different sizes of pottery can also be calculated. A small-necked vessel probably stored liquids or very small seeds, not large seeds. Large open vessels, such as bowls, probably weren't used for storage, since they would be difficult to seal from moisture, rodents, and bugs.

Problem 16. Calculate the circumference of the pot presented by the sherd drawn below.

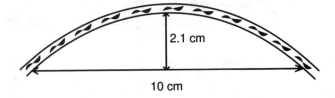

2.1 cm

10 cm

FIGURE 4
Sherd drawing

Pollen Analysis

Pollen, the male genetic substance produced by flowering plants, offers archaeologists a remarkable tool for studying the past. If the environmental conditions are fairly constant, pollen may be preserved in sediments over great time spans. Just as every kind of plant has a different kind of flower, pollens are also distinct, usually to the genus level.

Pollen is deposited in sediments by settling out of the air onto the ground or water surface. It also remains where people have processed plants, in areas such as hearths, kitchen areas, or storage rooms. Over time, pollen becomes part of the buried deposits.

Buried pollen can be recovered in several ways. Sometimes palynologists (specialists who study pollen) will study a core sample taken from a lake bed, swamp, bog, or ocean floor, along with samples from arroyo banks or pack-rat nests. Archaeologists send them samples from ancient sites to be analayzed, also. The characteristic all of these situations have in common is that the top of the core, arroyo, pack-rat nest, or archaeological site is the most recent deposit, and the bottom is the most ancient.

Why study pollen? Archaeologists use it to understand two main topics: how ancient people used various plants, and what past environments were like. Pollen analysis can tell us if people grew their own food, relied on wild plants, or some combination of both. It is also possible to learn about past climates, because every plant species has specific requirements for temperature and moisture. The climate necessary to support the plants represented in the pollen samples can be inferred.

By studying a sequence, such as the entire lake bed core or samples from every level of an archaeological site, we can observe change over time. Climatic and environmental changes, as well as human cultural change, can be learned. Pollen analysis gives us a very dynamic window on the past.

Once in the laboratory, pollen is recovered from sediments through a complex procedure termed "pollen extraction," which essentially involves dissolving the sediments while retaining the pollen. The vial of recovered pollen is in an oil suspension, and a small amount of it is placed on a microscope slide. Magnifying the grains 400 to 1000 times, palynologists count and identify the pollen. They are then ready to begin interpreting the results, assisted by statistical analysis.

Identifying and counting each of the pollen grains on a single slide can be quite tedious because there are typically many thousands of grains. Remember, the slide contains only a small amount of the pollen in the whole sample, and the counting procedure does not tell the analyst how much of the sample is represented on the slide. Even if every grain of the sample is identified and counted, some questions simply cannot be answered from such a procedure, mainly having to do with the density of pollen within the sediments. It can also be useful to know how long it took for a unit of sediment to be deposited in a certain area. This knowledge gives insight into past climatic processes and also provides a basis for comparison between samples.

The solution to both the tedium of counting a whole slide and the density-related questions is to introduce a known quantity of foreign grains to the pollen sample. Lycopodium spores are often used, added in tablet form during the pollen extraction procedure. While counting, the analyst records the number of Lycopodium spores, as well as the numbers and types of pollen grains. Since the density of the Lycopodium spores is known, the density of the various types of pollen grains can be determined.

For example, if the analyst records 400 grains of pine pollen and 800 Lycopodium spores, then (assuming the spores were evenly mixed into the sample) there must be half as many grains of pine pollen as the number of Lycopodium spores that were added. You can calculate this "total in sample" number for the pine pollen by equating the proportions. Say there were 10,000 Lycopodium spores added to the sample. Then:

$$\frac{800 \text{ spores counted}}{10{,}000 \text{ spores added}} = \frac{400 \text{ pine grains counted}}{x \text{ total pine grains in sample}}$$

$$\frac{800}{10{,}000} = \frac{400}{x}$$

$$800x = 400 * 10{,}000$$

$$x = \frac{400 * 10{,}000}{800} = \frac{4{,}000{,}000}{800} = 5{,}000$$

So there were a total of 5,000 grains of pine pollen in the sample.

Analysts sometimes count up to 1000 grains of pollen to get a valid sample. The goal is to count enough pollen to assure a relative representation of the plants growing in the vicinity, including the more rare plants.

To find out how much pine pollen there is compared to other kinds of pollen in the sample, calculate the percentage of the total grains of ALL types of pollen. Say there were 1,000 grains of pollen of all types counted in the sample, and 400 of them were pine.

"% of total" is calculated by dividing the pine pollen count by the total count:

$$\frac{400 \text{ pine pollen}}{1{,}000 \text{ any pollen grains}} = \frac{400}{1{,}000} = 40\%$$

So 40 percent of the pollen grains in the sample were pine. 60 percent must have been other sorts of pollen.

Problem 17. In this example, our analyst has processed two samples of identical size, and has counted and identified 1000 grains of pollen for each.

Sample A comes from a level of 1000-year-old sediment just below the first occupation of a North American village site; it will tell us the kind of environment the first villagers lived in. Ten thousand Lycopodium spores were added to the sample. Below is what the analyst found on the slide:

Plant Pollen	Counts	Total in Sample	% of Total
Pine	300		
Sagebrush	226		
Grass	275		
Oak	72		
Spruce	100		
Cactus	27		
Total	1000		100

Lycopodium spores = 470

Sample B comes from 900-year-old sediments covering the village site; it will tell us the kind of environment at about the time the then 100-year-old village was abandoned. Again, 10,000 Lycopodium spores were added to the sample.

Plant Pollen	Counts	Total in Sample	% of Total
Pine	290		
Sagebrush	240		
Grass	300		
Oak	60		
Spruce	90		
Cactus	20		
Total	1000		100

Lycopodium spores = 720

Calculate the "total in sample" and "% of total" for both tables.

Compare the two samples. First look at "% of total." Are the relative plant pollen grain percentages between Samples A and B significantly different? A bar or line graph can help make the comparison.

Now compare the "total in sample" column. The total for Sample B is what percentage of the total for Sample A?

Summary. By comparing the "total in sample" results of the two samples, it's clear there has been a major change in the plant environment near the village site over the 100-year time span. By comparing the "% of total" of the two samples, we can see that the change has not been one of plant community composition. In other words, the same plants in about the same percentages are still there.

If we had not measured pollen density, we would not be aware from the percentages that any change had occured. The palynologist and archaeologist would pose hypotheses to explain the decline in total pollen. One possibility is that the sedimentation rate increased over time. The unit of sediment that earlier took 100 years to accumulate (including 100 years-worth of pollen in it), may later take only 65 years to accumulate (including 65 years-worth of pollen). Increased erosion is one cause of this. Another possibility is that people cleared the nearby native vegetation for farmland or wood. Can you think of other reasons? How might you test your ideas?

Maryam Shayegan Hastings
Mathematics and Computer Science

I was born in Iran, and studied in the European-style school systems there, where mathematics is emphasized. My family immigrated to the United States while I was in seventh grade. Although I did not speak English, I was more advanced in mathematics than my classmates, so my teachers encouraged me to take the more challenging mathematics courses. In college, I majored in mathematics. After graduation, I worked for a year at Bell Laboratories as a computer programmer.

At this point, I realized that I wanted a career in education, which required a graduate degree. So I entered graduate school at the University of Michigan, where I found the climate in the Department of Mathematics very chilly for women. During my first year of graduate work, I married. My first son, Iraj, was born while I was pursuing a PhD in mathematics. Writing a dissertation while caring for a new baby was the most challenging task I had ever encountered. At that time, women in my situation received very little support. My husband and I shared in the care of our son, but it was still very difficult.

In 1975, I received a PhD from the University of Toledo and started my college teaching career. My second son, Ramin, was born in September 1981, the only time I was obliged to take a semester's leave. My colleagues were not very supportive, and I experienced the first setback in my teaching career. By the time my third and last child, Shirin, was born in 1983, I had changed jobs and did not have to interrupt my career. In fact, I was able to administer finals before she was born. After Shirin's birth, I returned to graduate school on a part-time basis to earn a master's degree in computer science.

For the past nineteen years, I have been a teacher. Presently, I am Professor of Mathematics and Computer Science, and Chair of the Department of Mathematics/Computer Science/Information Systems at Marymount College in Tarrytown, New York. Marymount is a women's college, and I have a strong interest in promoting mathematics for women. I speak to secondary school audiences as a participant in Women And Mathematics (WAM), and direct a summer program in mathematics and science for high-school girls at the college.

My study of the programming language C, described in one of the problems below, led to my writing a C manual that accompanies a textbook in programming languages written by two of my colleagues at Marymount college, Doris Appleby and Jay VandeKopple. My manual was published by McGraw Hill, and I am currenty working on the second edition.

Aside from the family support I have enjoyed, my career has been made possible—and rewarding—by the support and encouragement I have received from my colleagues. It is very important for young people to create networks of friends to help them through the challenging times that accompany any exciting work.

Simple Closed Curves

My mathematical area of study is topology. In geometry, most students learn that a square and a triangle are different geometric forms. However, to a topologist these two forms are equivalent. It's easy to create a square or a triangle from the same rubber band by simply stretching it. The way a topologist looks at it, if one shape can be elastically twisted and stretched into the same shape as the other, then these two shapes are topologically equivalent.

All the curves in Figure 5 are topologically equivalent, but they are not equivalent to any of the curves in Figure 6. The curves in Figure 6 are not topologically equivalent to each other, either.

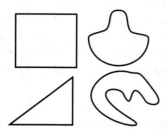

FIGURE 5

Topologically equivalent curves

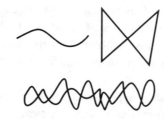

FIGURE 6

Topologically different curves

All the curves in Figure 5 are called simple closed curves. A curve is closed if all its end points are connected. If you start anywhere on the curve and trace all of it without retracing any segment, you end up where you started. If it never crosses itself, it is a simple curve. None of the curves in Figure 6 is a simple closed curve. The Jordan curve theorem states that a simple closed curve divides the plane into two regions, one inside and one outside the curve.

Problem 18. Are the points A, B, and C in Figure 7 inside or outside the simple closed curves?

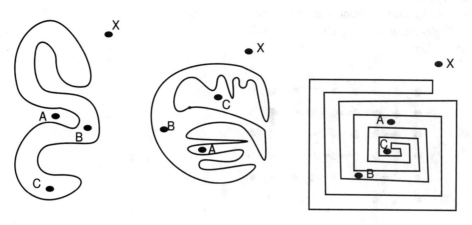

FIGURE 7

Digital Topology on a Computer Screen

Although most of my training has been in mathematics, I find the study of computer science fascinating. For a number of years, I have been meeting weekly with a group of topologists in the New York area. Some of these mathematicians became interested in the topological study of the computer screen. It is very different from the plane, since it is defined by a finite number of pixels. Forms appear on the computer screen, depending on which pixels are turned on or off. The study of the computer screen is an application of digital topology. This is an area of research that connects the two fields of topology and computer science by providing a theoretical foundation for many of the common operations in computer graphics.

Three of the topologists in the group, E. Khalimsky, R. Kopperman, and P. Meyer, discovered a notion called Connected Ordered Topological Spaces (COTS). Using COTS, they created a single definition of connectedness. As my contribution to their work, I wrote computer programs in C to illustrate some of their theoretical work. For example, one of my programs will accept as input a set of points on the screen and determine if they make a simple closed curve. Given a simple closed curve and a point, the program determines if the point is inside or outside the curve.

There are many applications for such research. We depend on computer graphics for solutions to problems. For instance, physicians rely on computer-generated pictures to diagnose and cure illnesses. As you might imagine, it is very important for a brain surgeon to know if a tumor appearing on a computer-generated picture is inside or outside of the brain.

How do we define a simple closed curve on the computer screen? What does it mean to say one point is next to another point? This concept of being "next to" is called adjacent. There are two definitions of this concept: four-adjacency and eight-adjacency. In a four-connected curve, the four points immediately above, below, left, and right of a given point are adjacent to the point. In an eight-connected curve, the four neighbors diagonal to the given point are also adjacent. L. N. Stout's (a mathematician) definition of a simple closed curve on the computer screen is, "Each point on the curve must be exactly adjacent to two other points on the curve."

Problem 19

Which of the curves below are:
  Four-connected simple closed curves?
  Eight-connected simple closed curves?
  Neither?

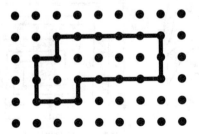

FIGURE 8

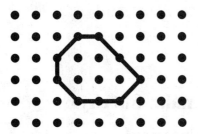

FIGURE 9

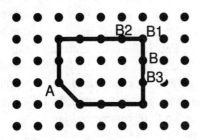

FIGURE 10

Perceived Authority and the Game Show Controversy

The following problem was posed by *Parade Magazine* columnist Marilyn Vos Savant in the February 17, 1991 issue. Her article is very entertaining, well worth going to a public library to look up.

She had posed the problem and the answer to it in a previous issue. But nine out of ten readers completely disagreed with her solution. These readers included mathematicians from prestigious universities as well as the general public. She finally reported that a PhD from MIT confirmed her solution.

Problem 20. "Suppose you're on a game show, and you're given a choice of three doors. Behind one door is a car; behind the others, goats. You pick a door—say, number one—and the host, who knows what is behind all the doors, opens another door—say, number three—where there is a goat. He then says, 'Do you want to pick door number two?' Is it to your advantage to switch your choice?"

Donna McConnaha Sheehy
Civil Engineering

Becoming an engineer was not something I planned, it happened strictly by accident. I was born and raised in a small town in Montana, where young women were expected to become homemakers, secretaries, or teachers. None of those areas interested me—I wanted to be a great artist. I spent my high school years preparing for a career in the commercial art field, taking art, speech, journalism, and mechanical drawing classes. In fact, I was the first female student to take mechanical drawing, a major breakthrough in those days. Because I was concentrating on a liberal arts background, my last math class was in my sophomore year. In 1971, I graduated from Anaconda High School, ready to conquer the world of commercial art.

My art training started at the College of Great Falls, in Great Falls, Montana. I spent one miserable year there. I didn't like being told when to draw, who to draw, and what style to imitate. Art was something I had enjoyed, but it was no longer enjoyable.

The summer of 1972 found me at home, looking for a job so I could afford to go back to school, even though now, I didn't know what I was going to major in. My mechanical drawing background got me a work-study draftsperson position for the Forest Service in Butte, Montana. That was my first exposure to the world of engineering. Short finances forced me to reconsider my choice of colleges, and I decided to stay at home and attend the local university, Montana College of Mineral Science and Technology. The work I had done as a draftsperson piqued my interest, so I decided to try engineering as a major.

My first year as an engineering student was tough. I had taken no math classes since my sophomore year in high school and I was lost! I had to go back to basic math, algebra and trig. As a result, it took an additional year to obtain my engineering degree.

While in school, I worked part-time for the Forest Service, and full-time each summer. The job became a training position, and upon graduation in 1977 I went to work as a Civil Engineer for the Deerlodge National Forest in Butte, Montana. Marriage moved me away from Butte in 1978, and for the next 12 years I worked for the Helena National Forest in Helena, Montana. In 1990, I began working for the Northern Region Forest Service in Missoula, Montana.

My responsibilities have varied throughout my career, from road location, layout, design, contract preparation, and construction inspection to facilities, bridges, transportation planning, logging engineering, computer systems, mapping, surveying, and supervisory responsibilities. Currently, I am responsible for road management of the thirteen national forests in Region One, Northern Region. This includes signing, highway safety, maintenance, and various road management programs. I juggle my career with the responsibilities of motherhood. My son was born May 12, 1992, and he keeps my husband and me very busy.

Network Analysis

The Forest Service is planning a resource project in Section 21. The area is primarily surrounded by two private ranches: Ranch A and Ranch B. There is an existing road into the area, but the Forest Service does not have legal rights to use the road, so the government will have to obtain a right-of-way from each ranch. The existing road cannot be used in its present condition and will have to be reconstructed. Reconnaissance of Section 20 indicates potential roading problems associated with wet springs, weak soils, and rock outcrops. See Figure 11.

There are two possible alternatives to providing access to Section 21:

Alternative 1 will utilize the existing road across Ranches A and B. Right-of-way will be needed, at an average cost of $20,000 for each case. Reconstruction is needed, and the costs will vary by segment.

Segment	Miles	$Cost/Mile
1-4	0.2	$5,000
4-5	0.5	$15,000
5-6	0.3	$10,000
6-3	0.3	$12,000

Six inches of surfacing is needed for segment 1-4-5-6-3, at a cost of $10,000/mile. Cattleguards will be needed at points 4, 5, and 6 at a cost of $3,200 each.

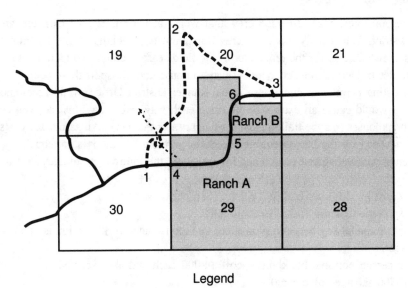

Legend

Existing Road	———————
Proposed Road	- - - - - - - - -
Private Land	▨▨▨▨▨
Road Segments	•———•
Stream	⌒‥‥‥‥

FIGURE 11
Sections and Roads

Alternative 2 will build a new road around the private land. Segment 1-2 will cross some of the problem areas in Section 20. This will require more drainage structures and surfacing than Segment 2-3 will need. Segment 1-2 is 1.3 miles long, and will cost $25,000 per mile to construct. A major drainage structure will be needed to cross a stream in Section 30, at a cost of $7000. Nine inches of surfacing is needed at a cost of $15,000 per mile.

Segment 2-3 is 1.0 miles and will cost $17,500 per mile to construct. Six inches of surfacing is needed at a cost of $10,000 per mile.

Problem 21.   For the resource project, determine the most cost-effective access route with the least environmental impact. Is it beneficial to avoid the right-of-way acquisitions?

Managing Work Crew Hours

You are responsible for 640 person-hours of work that will be done this season in a remote location. Your crew must walk in and out each day, an average of four hours per day (two hours each way). Traditionally, you have used a four-person crew that costs $18.00 per hour.

Problem 22.   Is it more economical to work your crew a normal eight-hour day, or should you ask them to work ten-hour days and pay time-and-a-half for overtime?

Locating Work Crews

You are responsible for a six-person crew that works at various remote locations around the Ranger District. You usually work the crew for about two months at one location and then move it somewhere else for the remainder of the field season (about two months). One of your tasks is to determine the most economical way to house and transport these people.

An existing permanent bunkhouse at the Ranger District Office would accommodate the crew, but it would cause an extra 40-mile drive each way, every day. Instead, you could buy portable bunkhouses (maybe trailers) and set them nearer the work. All associated costs (meals, bedding, etcetera) for the permanent bunkhouse or portable bunkhouses are equal.

Through your exhaustive search and investigation, the following estimates have been compiled:

Crew cost is $6.00 per hour per person ($36 per crew hour)

Transportation cost is $0.25 per crew-mile

Travel time one-way between permanent bunkhouse and work location is one hour (average)

Three-person portable bunkhouses cost $4,000 each and are expected to last eight years ($300 salvage value each)

Extra maintenance for portable bunkhouses is $200 each

Crew will move the portable bunkhouses

Spring set-up and move-out, and fall move-in and winterize will take eight crew-hours each move.

One midseason move (average) takes eight crew-hours

The field season averages four months each year (88 work days)

Interest rate is 8 percent

Problem 23.   Should you buy portable bunkhouses, or use the existing permanent bunkhouse at the Ranger District Office to accommodate the crew? For each option compute the cost per work hour.

   If portable bunkhouses are purchased, the present value of the salvage value of the bunkhouses should be subtracted from the $8,000 cash purchase price. Then, assume that 8 equal payments will be made at the *end* of each year to pay off the debt.

Transportation System Planning

You are responsible for transportation system planning on a timber-harvesting project that will enter a previously unroaded drainage. The topography in this area is suited to either a valley bottom road with downhill yarding, or ridgetop roads with uphill yarding. Preliminary estimates of construction costs indicate that the valley road will cost $55,000 while the climbing road to the ridge tops will cost $94,000. Uphill cable yarding costs are $35/mbf and downhill yarding costs are $60/mbf (mbf = mille or thousand board feet). Volume to be harvested from this area is two million board feet.

Problem 24.   Which road would you recommend be built? What is the break-even volume?

Length of a Tree

The angles in Figure 12 are measured with a clinometer, which measures degrees.

Problem 25   What is the length of the tree shown in Figure 12?

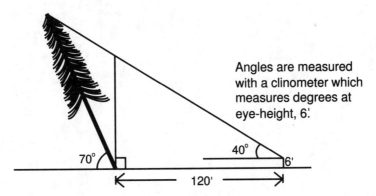

Angles are measured with a clinometer which measures degrees at eye-height, 6.'

40°

6'

70°

120'

FIGURE 12
Length of Tilted Tree

Guyline and Skyline

When the ground is level enough, timber can be harvested with a tractor. If the ground is too steep (greater than 40%), it is necessary to build a skyline for a cable suspension system (using a block, or pulley wheel) that can haul the logs above the ground, similar to a cable car or aerial tram.

To calculate how many logs can be suspended at one time, it is necessary to know the vertical lift of the skyline. Too many logs will break the skyline or pull down the tree that suspends the skyline. Not enough logs will make the harvesting process take longer than necessary and increase costs.

Problem 26. Given a tension exerted on the skyline by the yarder of 55,000 pounds, and assuming that the skyline is straight:

A. Determine the log load that can be lifted by the skyline.

B. Determine the tension in the guyline and the vertical load acting on the tree.

(These tensions are used to determine proper cable and anchor strengths.)

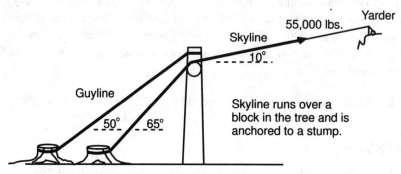

FIGURE 13
Guylines and Skyline

Linda Valdés

Mathematics

When I was growing up, women were not expected to choose a career in mathematics. Since I was always a rather hard-headed, stubborn individual, I rebelled against this notion and decided to do it anyway. And now, as a mathematician, I derive great satisfaction at the surprise in people's faces when they find out what I do for a living.

Mathematics is great. I can sit and think, deduce and reason, be creative all day long—and get paid for it. And what's really interesting is that most of the work I do has no immediate practical value, because mathematicians don't usually attack a problem with an application in mind. We leave it to the scientists to take our results and find some use for them.

As for my history: I majored in mathematics in college and obtained my BA from the University of Florida. Next, I joined the Peace Corps and went to Liberia, West Africa, where I had my first experience as a teacher. I taught mathematics, and my students taught me about their culture and what it meant to be an African. After my tour in Africa, I went to Scotland to teach math, and found that mathematics is important in all cultures.

Upon returning to the United States, I went back to school, earning a BA in art at the University of California, Santa Cruz (UCSC). As I sometimes say, "art is for my heart, math is for my head." I continued teaching math in high school, finally returning to UCSC to obtain a master's degree and PhD in mathematics. I am now an assistant professor at San Jose State University where I teach math and research all day long. It's a lot of fun.

The problem I have included is one I was fiddling with awhile back. At the time, I was working on the bigger question of finding the graphs with the fewest number of spanning trees. The following graphs were the very ones I needed.

Graphs and Spanning Trees

A graph is a set of points or vertices that may or may not be connected by edges. These structures are studied in the field of mathematics called graph theory. They are used to help solve problems in computer science, combinatorics, electrical engineering, game theory, and many other areas. The following problems are similar to those that might be faced by a designer of computer hardware or computer networks.

If edge e joins vertex v to some other vertex in the graph, we say they are adjacent. In a cubic graph, every vertex is adjacent to exactly three edges. (See Figure 14.)

Think of the vertices as members of a school club, and each edge as phone calls made between members. Figure 14 might describe a club with four members, and the three phone calls each member made to the other members. In a computer network, each vertex is a place where information is gathered and dispersed, and the edges are the wires that carry the electrical impulses.

Network hardware designers consider both efficiency and cost in their designs. They like to make sure all the vertices are reachable from each other.

Problem 27. Six cities communicate with each other through telephone lines. There are exactly three lines running out of any one city. How many ways can cities A, B, C, D, E, and F be connected with telephone lines? Try drawing the cities in order around a circle and connecting them to each other.

Problem 28. In Figure 14, a "cycle" can be traced by starting at the vertex v_0, traveling across edge e_0 to v_1, traveling across e_1 to v_2, and finally traveling across e_2 to v_0, the starting point. No vertex is encountered more than once along the cycle, except v_0. Think of this cycle as describing phone calls made from one member of the school club to the other three.

A tree is a graph with no cycles. A spanning tree of a graph (G) is a subgraph of G. It is a tree and includes every vertex of G. One spanning tree of the cubic graph in Figure 14 is shown in Figure 15. There are 15 more spanning trees in the graph. Try to find them. Spanning trees are important in networks because every vertex is reachable from every other vertex with a minimal amount of wiring.

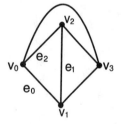

FIGURE 14

A cubic graph

FIGURE 15

A spanning tree of the cubic graph

Problem 29. A family of graphs is a set of graphs with a common set of properties. The set of cubic graphs shown in Figure 17 is a family of cubic graphs with the property of containing a number of subgraphs (such as Figure 16) that are connected by edges. Each graph in Figure 17 contains one or more subgraphs like the one in Figure 16, connected by edges.

The last example in Figure 17 is a generalization of the family, where dots may be replaced by more subgraphs such as the one above. How many spanning trees are there in each graph of the family?

Hint: Let p = number of vertices in the graph. Think of the graph as a necklace with each bead as a subgraph in Figure 16, and each edge connecting two subgraphs, like a piece of string.

FIGURE 16
Subgraph

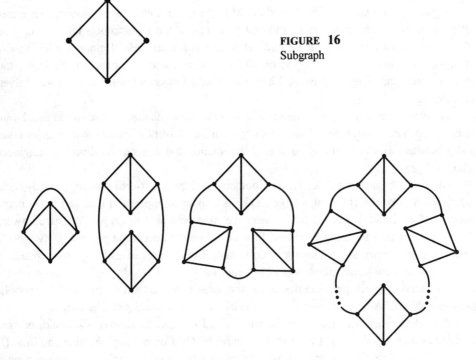

FIGURE 17
Family of Graphs

Jill S. Baylor

Electrical Engineering

I grew up in a family of four children. My father had a PhD in engineering. He expected all of us to do well in school and go to college at one of the state schools in Virginia. I really liked math courses and took all I could get until I graduated from Hampton High School in Hampton, Virginia. However, I really didn't know what career I wanted, so although I took a lot of science (including chemistry), I had not yet taken physics when I entered the University of Virginia.

Halfway through my first semester in a liberal arts curriculum, I discovered that I wanted to be an engineer. Fortunately, I was already taking a lot of physics and mathematics courses, and graduated with a degree in Applied Mathematics and a minor in Electrical Engineering after four years.

My first job was with Duke Power Company in Charlotte, North Carolina, mostly because my husband-to-be (I met him at college and he is also an engineer) and I wanted to get married, but only if we could live in the same place. We were both able to get jobs at Duke Power.

There, I was a Planning Engineer, focused on power generation planning. My work emphasized two primary areas: looking at costs and efficient uses of existing power plants; and deciding when new power plants would be required.

We worked on our graduate degrees at the same time, and after a lot of night school, we both received MBA degrees from the University of North Carolina at Charlotte.

After five years with Duke Power, we decided to move to Denver, Colorado, where the energy business was thriving. I went to work in Mobil Oil Corporation's Mining and Coal Division. At Mobil, I studied the coal markets, examining the costs of mining and transporting coal to the primary customers—electric utilities. During this time, I became a registered professional engineer in Colorado.

By 1983, the energy business was not doing so well. So the next year, I got a job at Stone & Webster Management Consultants that utilized the knowledge and skills I had gained at Duke Power and Mobil. At Stone & Webster, I provided consulting services to electric utilities and others that were building power plants. I concentrated on generation and transmission planning, and fuels-related work.

In 1992 I decided it was time for another job change and I went to work for a firm called RCG/Hagler Bailly in Boulder, Colorado as a principal, managing the utility planning practice. I am doing the same kind of work that I did at Stone & Webster, but I also manage staff and have more responsibility. I still use my math skills every day, and consider myself fortunate to have found a career that suits me so well.

Selection of Lowest-Cost Coal

Efficiency measures how much fuel is required to produce electricity.

One Megawatt = 1,000 kilowatts

A 100 watt light bulb will consume 100 watthours if it is turned on for one hour. Ten of these bulbs will consume 1,000 watthours in one hour or one kilowatthour.

One ton = 2,000 pounds

Btu = British thermal unit

Problem 30. An electric utility in Kansas needs coal to fuel one of its power plants, so it requests bids from coal suppliers. Two suppliers in Wyoming respond. One supplier offers to provide 8,400 Btu-per-pound coal for $14 per ton delivered to the power plant. The second coal supplier offers to provide 8,800 Btu-per-pound coal for $17 per ton delivered to the power plant. If the power plant burns either coal with an efficiency of 10,500 Btu-per-kilowatthour (kWh)—its so-called heat rate—which coal is cheaper on a dollar-per-Megawatthour (MWh) basis for the utility to burn?

Determination of Need for New Electricity Generation

One Megawatt = 1000 kilowatts

A 100 watt light bulb will consume 100 watthours if it is turned on for one hour. Ten of these bulbs will consume 1,000 watthours in an hour or one kilowatthour.

$$\text{Reserve margin (\%)} = \frac{100 * (\text{Installed Capacity} - \text{Peak Demand})}{\text{Peak Demand}}$$

Problem 31. A Midwest utility currently has four power plants generating electricity. Unit A is oil-fired and can produce 100 Megawatts (MW). Unit B is fired with natural gas and can produce 50 MW. Unit C is coal-fired and can produce 200 MW. Unit D is hydroelectric and can produce 25 MW.

The maximum amount of electricity needed by customers (the peak demand) has reached a level of 300 MW. Peak demand is growing at a rate of two percent per year. If the utility must keep a reserve margin of 15 percent, how many years from now will a new generating unit have to be installed?

Present Survey Results

Numbers representing raw data are sometimes hard to understand. It is usually easier to understand data when it is presented in percentages, rather than in actual numbers. Graphs that represent the data help in understanding, too.

Problem 32. A recent survey of electric utilities across the United States asked for information about generating units that are 30 years of age or older. Information was provided for 172 generating units. Of those 172, two were coal/gas, 90 were gas/oil-fired units, 34 were coal units, 26 were oil-fired units, and 20 were gas turbines. Create a pie chart demonstrating the percentage for each type of generating unit.

How Much Can a Generating Unit Produce?

One Megawatt = 1,000 kilowatts

A 100 watt light bulb will consume 100 watthours if it is turned on for one hour. Ten of these bulbs will consume 1,000 watthours in an hour or one kilowatthour.

There are 365 days in a non-leap year. Each day contains 24 hours for a total of 8,760 hours each year.

Problem 33. A 300 Megawatt (MW) coal-fired generating unit produces power 75 percent of the hours each year at a level of 300 MW (its full load). How much energy (Megawatthours) of electricity does it produce?

Calculate a Unit's Capacity Factor

1 Megawatt = 1,000 kilowatts

A 100 watt light bulb will consume 100 watthours if it is turned on for one hour. Ten of these light bulbs will consume 1,000 watthours in an hour or one kilowatthour.

There are 365 days in a non-leap year. Each day contains 24 hours for a total of 8,760 hours each year.

Capacity factor = the amount of energy produced, divided by the maximum amount possible over a period of time.

Problem 34. A generating unit is capable of producing 200 Megawatts (MW) in each hour of the year. If it generates 1,103,760 Megawatthours (MWh), at what capacity factor is it operating?

Costs for Purchasing Power

One Megawatt = 1,000 kilowatts

A 100 watt light bulb will consume 100 watthours if it is turned on for one hour. Ten of these bulbs will consume 1,000 watthours in an hour or one kilowatthour.

Problem 35. Part A. Utility B agrees to buy power from Utility A. The arrangement says that Utility B will pay for 100 Megawatts (MW) at the rate of $5 per kilowatt for each month ($5/kW-month). This is called a capacity charge. In addition, for each kilowatthour purchased, Utility B will pay $.02 (two cents/kWh). This is called an energy charge. If Utility B buys 569,400 megawatthour (MWh) in one year, how much will Utility B pay for both capacity and energy during a year?

Part B. A utility determines that in order to meet its obligations to stockholders and financial institutions next year, it must charge customers $145,920,000. If all the customers use 1,824,000 Megawatthours (MWh), what is the average rate in dollars per kilowatthour ($/kWh) the utility will charge its customers?

Part C. A utility plans to build a 345-kilovolt transmission line that costs $750,000 per mile. Forty-six miles of line will be built. How much will this project cost?

Selection of Coal Transportation Method

Moving coal by railroad costs 2¢ for each ton-mile. Moving coal by truck costs 15¢ per ton-mile. Moving coal down waterways on a barge costs 1¢ per ton-mile.

Problem 36.   Which method of moving coal costs the least?
  1) 100 miles by truck
  2) 900 miles by railroad
  3) 1000 miles, of which 250 miles is by barge and 750 miles is by railroad.
Assume that equal tonnage is moved for each alternative.

Clean Air Impacts on Coal

  One ton = 2,000 pounds
  Btu= British thermal unit
  Sulfur has an atomic weight of 32. Oxygen has an atomic weight of 16. Thus SO_2 has an atomic weight of 64 or twice the atomic weight of elemental sulfur. For each pound of sulfur in the coal, two pounds of SO_2 are generated when the coal is burned.

Problem 37.   The coal that will be burnt at Utility A's power plant has a heating value of 12,000 Btu/pound and a sulfur content of one percent.

  Under the Clean Air Act, Utility A must limit levels of sulfur dioxide (SO_2) to no more than 1.2 pounds SO_2/MMBtu. What percentage of the SO_2 emissions will have to be scrubbed out with flue gas desulphurization to comply with the law?

Lynn R. Cominsky

Physics; X-ray Astronomy Research

After graduating in 1971 from Sweet Home Senior High School in suburban Buffalo, New York (where I took four years of math, including calculus), I attended Brandeis University near Boston, Massachusetts. Originally, I planned to become a psychology major; however, my first class quickly changed my mind! I had been a vegetarian since the age of 16, and most of the class dealt with dissecting cat's brains to figure out how their vision worked. I decided I was too squeamish for such work (and for medicine) and changed my major to chemistry.

So I spent the next three years as an undergraduate in physical chemistry research. It was exciting work, but very painstaking. A small mistake (e.g., a spilled solution or incorrect measurement) would mean redoing the entire day's work.

At the same time, Brandeis had just entered the field of computer science, purchasing its first mainframe computer and offering computer classes. I became interested in computers because there were many job opportunities available on campus. I also liked the fact that, if I made a mistake, I could just correct the program and continue with my work, without losing all that time and effort.

In those days, few women were majoring in chemistry or computer science, and Brandeis had no women professors for either of these subjects. The computer courses counted towards a physics major, as did many of my chemistry courses, so I ended up with a double major in physics and chemistry.

I had done poorly in my first physics course at Brandeis (probably because I had never taken high school physics), and didn't care much for the field. But during the last physics course, Modern Physics, that was required for my double major, I fell in love with the subject. Quantum mechanics and relativity were much more interesting than the problems involved in first-year physics (balls rolling down ramps, pulleys, and the like). The Brandeis physics faculty were very supportive of my efforts, so I continued with physics. In 1975, I graduated from Brandeis magna cum laude, earning a BA in Physics with Honors in Chemistry.

Following graduation, I looked for jobs in both chemistry and computer-related areas, and was fortunate to be hired at the Harvard-Smithsonian Center For Astrophysics (CFA), analyzing data from the UHURU X-ray satellite. It was the first satellite to observe x-rays from space. X-rays are produced by very unusual stars such as white dwarfs, neutron stars, and black holes, usually found in binary systems with stars similar to our Sun.

At CFA, I finally met a woman scientist, Dr. Christine Jones. She became my first mentor. Christine and her husband, Dr. William Forman, were my supervisors on the UHURU data analysis project. Christine was also primarily responsible for my decision to attend graduate school and continue my research in the field of x-ray astronomy. To this day, her encouragement, advice, and enthusiasm influence my life. She continues to work with many other women college graduates employed in similar positions at CFA.

In 1977, I entered graduate school, studying physics at the Massachusetts Institute of Technology (MIT). There I discovered a neutron star pulsar (one of only 15 known at that time), using data from the SAS-3 satellite. A pulsar is a neutron star that rotates while sending out beams of x-ray light, similar to the beams sent by a lighthouse.

Although I expected MIT's environment to be less friendly than CFA's, in reality there are many women employed in the physics department, both as faculty and in important administrative positions. They actively recruited me and organized many events for new women students. There were no other women graduate students in my immediate research group, although there were always some in my classes. I received my PhD in Physics from MIT in 1981, after studying bursts of x-rays from neutron stars with Professors Walter Lewin and Paul Joss. I also met my husband, Dr. J. Garrett Jernigan, in the SAS-3 research group. We moved to California when we were both offered positions at the U.C. Berkeley Space Sciences Laboratory (SSL).

For the first three years at SSL, I enjoyed a postdoctoral fellowship and pursued my own research. I also became involved in the Extreme UltraViolet Explorer (EUVE) satellite project. The project's mission was to conduct the first sky survey in this wavelength band, which is a little less energetic than x-rays. The scientific instruments for the EUVE were being built at SSL, but there was no way to quickly analyze the data.

I decided that EUVE should be controlled from SSL, as we had done with SAS-3 at MIT. So I wrote a proposal to NASA, including a detailed plan for this work. NASA agreed to let SSL control the scientific instruments on the satellite and to develop the ground data analysis system, and in 1984 I became the manager of the software part of the project. In 1985, I also took over the hardware part of it, in preparation for a shuttle launch in 1987.

Following the disastrous explosion of the space shuttle Challenger in 1986, I accepted a faculty position at Sonoma State University (SSU). I had previously enjoyed teaching a course there, and did not look forward to many more years as a manager. So when the opportunity came to change career paths, I eagerly accepted the position even though I'd be responsible for developing courses in electronics, an area where I had little direct experience.

Today, I teach electronics and physics. I conduct x-ray astronomy research for NASA, work that is supported by grants. I'm also the advisor to the SSU Society of Physics Students, which built a radio interferometer telescope on the roof of the science building. I am the first and only woman faculty member in the Physics and Astronomy department (out of seven permanent faculty), but we often have part-time women lecturers.

SSU is a very supportive atmosphere for women faculty in the sciences, and my work has been greatly appreciated and acknowledged. In 1991, I won the Excellence in Education award from the local Chamber of Commerce. In 1992, I was named SSU's Outstanding Professor. And in 1993, the Council for Advancement and Support of Education named me "The California Professor of the Year." I really enjoy the combination of teaching and research here, and feel lucky that I can be paid to do work that is so much fun!

This year I am on sabbatical at the Stanford Linear Accelerator Center, working with a group of particle physicists on a program in particle astrophysics. We are designing and building part of an x-ray astronomy experiment, to be launched on an Air Force satellite in 1995.

Observations of an Eclipsing Stellar System

The light from a star is being observed by an astronomer on Earth. The light that the astronomer sees during one 14-hour nighttime observation is shown in Figure 18.

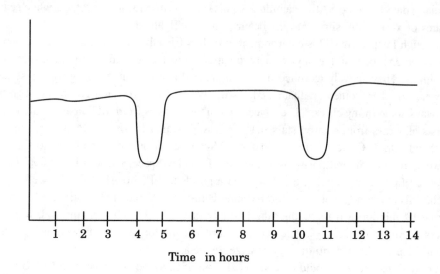

Time in hours

FIGURE 18

The astronomer decides that something (perhaps a planet like our Earth) is orbiting around the star (just as our Earth orbits around the Sun). As the planet circles the star, every so often it comes between the star and the Earth and blocks some of the starlight, preventing it from reaching the Earth. The planet and the star form a binary system, and the planet eclipses the starlight, as shown in Figure 19.

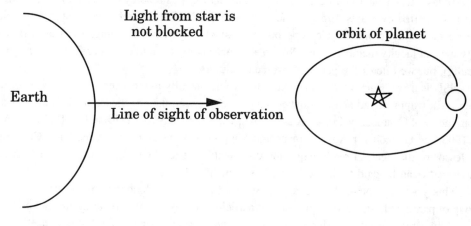

FIGURE 19

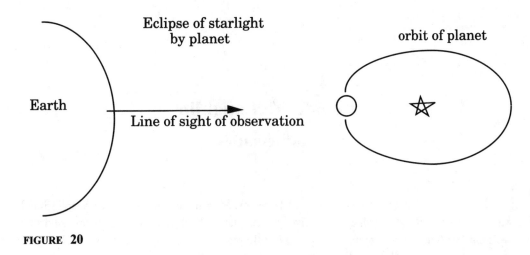

FIGURE 20

The famous scientist, Johannes Kepler, is responsible for a law of physics. It relates the orbital period of a binary system to the distance between the star and the planet. This law states that:

$$P^2 = d^3$$

where P is the orbital period measured in years, and d is the average distance measured in Astronomical Units (defined as the distance between the Earth and the Sun or approximately 1.496×10^8 km).

Problem 38. Assuming that the star and planet are being viewed from Earth as shown in the figures, can you help the astronomer figure out:

Part 1. How long it takes for the planet to orbit the star? (This is called the orbital period of the binary system.)

Part 2. How long the light from the star is blocked by the planet? (This is called the eclipse duration.)

Part 3. What is the orbital period for the binary system consisting of the Earth and the Sun? (How long does it take for the Earth to orbit around the Sun?)

Part 4. Given the orbital period for the star and the planet that you derived in part 1, how far apart are they?

Renate McLaughlin
Mathematics

I was born in what was then called East Germany. My parents fled to West Germany while I was still in elementary school, and I received the standard schooling for college-bound students: beginning with a foreign language (English) in fifth grade, adding a second language (Latin) in seventh grade, and adding a third language (French) in tenth grade. After completing thirteen years of school, students could apply to a college. Schools offered no choices whatsoever, all subjects were mandatory. Once a foreign language was started, students had to continue with it through the thirteenth grade, although the class would not necessarily meet every day of the week. Mathematics was mandatory in all thirteen grades, and the study of single-variable calculus was completed in high school.

One experience during my high-school years definitely had an impact on my later life—I spent a year as an exchange student at an American high school. At the time, I thought the study of various subjects was much more intense in the American high school than in Germany. On the other hand, I didn't think students could retain much of what they learned, because they had such a short exposure to the subjects.

German students finish the 13th grade with the equivalent of American general education requirements in a liberal arts college. When the time comes to apply to a university, students also decide on a major. I was debating between foreign languages, mathematics, physics, and music. One of my high school teachers spent many Sunday afternoons with me discussing the possible choices, and somehow we decided on mathematics. As a result, I entered the university in Münster, planning to study mathematics.

Three memories stand out from my years in Münster. To begin with, my first college mathematics class started with more than 200 students, but at the end of the second semester only 30 students remained. I felt terribly under-prepared, but somehow I managed to hang on. Second, there was the elderly professor who taught a course in numerical analysis. I was the only female student in his class, and he obviously did not approve of women studying mathematics. He would begin every lecture by looking directly at me and saying, "Good morning, gentlemen!" My third vivid memory concerns the exams arranged by the university in Münster. Students could elect to take special oral examinations at the end of a semester, and if they did well enough, they could receive a lower tuition rate for the following semester. Once, the senior professor initially refused to examine me, on the grounds that "a nice girl" did not belong in mathematics. I stood my ground, partly because I happened to know that the professor's daughter was also studying mathematics, and in the end he relented.

After completing the equivalent of a BA degree in mathematics at the university in Münster, I obtained a scholarship to The University of Michigan, and earned a PhD in mathematics there. I have taught at the University of Michigan–Flint since 1968 (family reasons dictated the location), and have been a Professor of Mathematics there since 1975. Unlike the typical routine at a large research university, I spend much of my time working and talking with undergraduate students outside of class, and I enjoy this contact with students very much. During sabbatical leaves I have taught at universities in Berlin, Germany, and Salzburg, Austria.

Lately, I have become interested in technological advances and how they change the way mathematics should be taught. My students use a graphing calculator in most of my classes, and I am learning to use computers effectively.

Somewhere along the way I realized a dream from way back: learning to fly and earning a pilot's license. By now, I have flown across the entire country in a four-seat airplane, most of the time as the only occupant of the plane. Many of the talks I give to high-school audiences are based on the mathematics a pilot must be familiar with to pilot an aircraft. A large airline crew has specialists for this, but a pilot of a private aircraft must know how to do it all.

Math Used by Pilots

Like mathematics, flying is on the borderline between science and art, incorporating a bit of each. In both flying and mathematics, there are many skills to be mastered, many principles to be understood, but in the end, the difference between "very good" and "superb" depends on intuition and experience.

A feeling for numbers is certainly helpful to a pilot. For example, suppose you are flying along on a heading of 315 degrees (remember that the compass rose is divided into 360 degrees), and air traffic control asks you to make an immediate turn to a heading of 170 degrees (perhaps because you are on a collision course with another aircraft). Unless you are specifically asked to turn right or left, you are supposed to turn to the new heading through the smaller of the two possible angles. Should you turn right or left? If you are on a heading of 315° and turn right to 170°, you will turn through an angle of 215°. But if you turn left to a heading of 170°, you will turn through an angle of only 145°. Therefore, you need to make an immediate turn to the left.

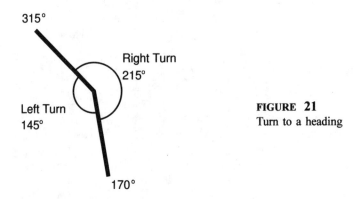

FIGURE 21
Turn to a heading

Many other situations call for a quick estimate involving numbers. For example, when there are air traffic delays due to bad weather or some condition at an airport, airplanes are stacked up in holding patterns. There are only three official ways to enter a holding pattern. The entry you choose depends on the angle between the heading of the aircraft as it approaches the holding pattern, and the layout of the pattern (printed in the packet of charts carried around by a pilot).

Now, look at some procedures that involve more than estimating differences of numbers. Four forces act on every airplane: thrust, drag, lift, and weight.

Thrust is created by the engine and pulls the airplane forward. Drag consists of *parasite drag* (friction, resistance from antennas and the shape of the wings), which increases as the plane moves faster, and *induced drag* (a by-product of lift), which is greatest at low speeds. Drag tries to prevent the plane from moving forward. Obviously, a plane must have more thrust than drag.

Lift is produced when enough air moves over the wing. (Bernoulli's principle explains this.) Lift tries to pull the airplane up. The opposite force— weight—pulls the plane down and

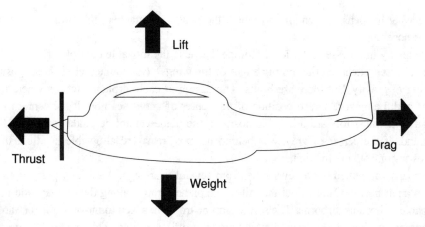

FIGURE 22
Four forces acting on an aircraft

is created by the weight of the passengers, cargo, fuel on board, engine, and the plane itself. For the plane to become airborne, it must be able to develop more lift than its weight.

The lift vector can be thought of as being attached to the plane at one specific point. This point—called the *center of lift*—is determined by the shape of the wings and how they are attached to the airplane. The center of lift does not change, regardless of whether the plane is empty or fully loaded. On the other hand, the center of gravity can change considerably, depending on the loading. You can observe this with an easily constructed model: Make a cross-section of a plane (from propeller to tail) out of thin cardboard. Put a small hole where the middle of the wings would be, pull a thread through the hole, and suspend the plane by this thread. This is your lift vector. Now attach a small weight to the plane from various places,

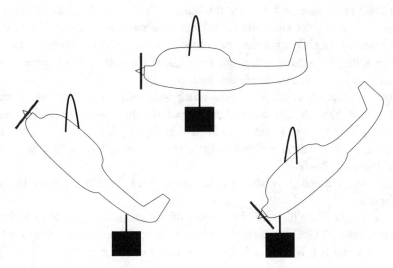

FIGURE 23
Effects of moving the center of gravity

and see what happens. Depending on where the weight is attached, the aircraft is nose-down, level, or nose-up.

Planes fly in an essentially level attitude, because the movable control surfaces on the tail (and, to a lesser extent, on the trailing edges of the wings), can counteract different positions of the center of gravity. But there are limits. The design of each airplane dictates where the most forward and the most rearward position of the center of gravity can be. If the center of gravity is ahead of its most forward allowable position, the plane cannot be landed and will dive into the ground. If the center of gravity is behind its most rearward allowable position, the plane becomes uncontrollable in flight.

Perhaps you remember a news story from several years ago. A military plane tried to take off from an airport in Canada, but the tail of the plane scraped along the runway, and the plane was unable to become airborne. The soldiers on board were asked to move as far forward inside the plane as possible, and a second attempted take-off succeeded. This is clearly a case where the pilot neglected to compute the center of gravity!

In summary, two things that make safe flight possible must be checked before every flight: the total weight of the loaded airplane, and the location of the center of gravity. Since the center of gravity is the total "moment" relative to some fixed reference plane—usually the front of the airplane or the firewall between the cockpit and the engine compartment—divided by the total weight, one can plot total moment on the horizontal axis, and total weight on the vertical axis, of a coordinate system. (See Figure 25.) The flight can be conducted only if the point lies inside the center of gravity/moment envelope.

Test pilots fly new aircraft under various weight and balance conditions to determine the "safe envelope," which is published as a graph in the operating handbook (Figure 25). The operating handbook for each airplane makes it easy to compute both the total weight and the total moment. Before taking off on each flight, the pilot must verify that the center of gravity is within the envelope.

To work these problems, it is not necessary to understand what "moment" is, but here is an explanation: The easiest way to describe "moment" is to think of a seesaw. For a heavy and a light child to balance on a seesaw, the heavier child must sit closer to the fulcrum of the seesaw, while the lighter child sits further out. The moment of each child is the child's weight multiplied by the child's position, or distance from the fulcrum. When the two moments are about the same, the seesaw will work best. So, if (heavier child * smaller distance) = (lighter child * greater distance), the seesaw balances.

In a plane, the moment of the passengers is their weight times their position, measured from the firewall. Rather than calculate the moment for each weight in the plane (fuel, baggage), it is easier to look up the values in a graph like the one in Figure 24, below. For more information on this subject, see the Federal Aviation Administration's *Pilot's Handbook of Aeronautical Knowledge,* 1980, pp. 75–78.

The following information is based on the operating handbook of a Cessna Cutlass RG, a four-seat airplane:

Given the weight of the fuel, read its moment off the graph in Figure 24 by looking at line 2. If the fuel weighs 200 lbs, its moment is about nine. If the rear passengers together weigh 300 lbs, their moment is about 24, as indicated by line 4 on the graph.

	Weight (lbs)	Moment (lb.-ins./1000)
1. Basic empty weight (includes unusable fuel and full oil)	1661	63.8
2. Usable fuel (at 6 lbs/gal)[maximum: 62 gal]		
3. Pilot and front passenger		
4. Rear passengers		
5. Baggage area 1 [maximum: 200 lbs]		
6. Baggage area 2 [maximum: 50 lbs]		
7. Ramp weight and moment	sum of numbers above	sum of numbers above
8. Fuel allowance for engine start and taxi	−8	−0.4
9. Take-off weight and moment (add lines 7 and 8) [maximum: 2650 lbs]		

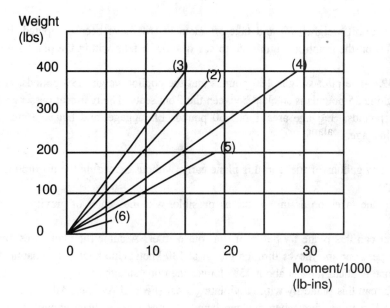

(2): Usable fuel   (3): Pilot and front passenger
(4): Rear passengers   (5): Baggage area 1
(6): Baggage area 2

FIGURE 24
Weight and moment graph

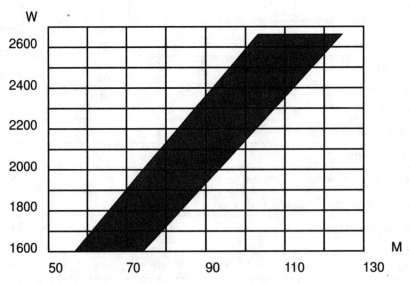

FIGURE 25
Center of gravity envelope

After take-off weight (W) and take-off moment (M) have been computed, look up the point (M, W) on the graph in Figure 25, to see whether it falls within the plane's envelope.

Problem 39. The pilot weighs 180 pounds and the copilot weighs 190 pounds. A packet of charts weighing 25 pounds is stacked between the front seats. The rear-seat passengers together weigh 340 pounds. Baggage area 1 has 60 pounds of luggage, and baggage area 2 has 15 pounds of luggage.

  a. How many gallons of fuel can this plane carry, without exceeding the maximum allowable weight?
  b. If the plane takes on as much fuel as possible, will the center of gravity remain within limits?
  c. How far can this plane fly before it runs out of fuel? Assume the plane uses nine gallons of fuel per hour and moves through the air at 130 knots. One knot = one nautical mile per hour; one nautical mile is about 15% longer than one statute mile.
  d. How far can this plane fly without violating FAA (Federal Aviation Administration) rules? The FAA requires that when a plane lands, it must have at least enough fuel to fly for another 30 minutes (more for certain flight conditions). If a plane lands with less fuel on board, the pilot's license may be revoked.

Problem 40. You have full fuel on board. You are the pilot and weigh 130 pounds. Today, you have one passenger who weighs 230 pounds. Baggage area 1 contains 150 pounds of luggage.

a. Can you take the passenger and all the luggage without exceeding your maximum allowable weight?

b. Can this passenger sit in either the front seat or the rear seat?

Problem 41. Three people are flying up north to go camping. They don't want to make fuel stops, so they fill the tanks to the top. The camping gear fills the baggage areas: 200 pounds in area 1 and 50 pounds in area 2. The pilot weighs 150 pounds. The two passengers (including a child) weigh 170 and 50 pounds, respectively.

a. Will this flight stay within legal limits as far as the weight is concerned?

b. Can the heavier passenger sit in front with the pilot?

c. Can the child sit in front with the pilot?

Problem 42. You have full fuel on board. You are the pilot and weigh 170 pounds. Supplies are being transported that weigh a total of 447 pounds. You would like the center of gravity to be as far back as is safe. (The plane operates more efficiently, and therefore uses less fuel, when the center of gravity is as far back as possible). How should you distribute the supplies?

Problem 43. You have only 44 gallons of fuel on board. You are the pilot and weigh 100 pounds. You are delivering 625 pounds of rather smelly supplies to another airport. To get away from the smell, you would like to put the supplies as far back as possible. How should the supplies be distributed?

Problem 44. You take off with full fuel. Two passengers in the front seats weigh a total of 360 pounds, two passengers in the back seats weigh a total of 235 pounds, and a picnic basket in baggage area 1 weighs 20 pounds. When you land, all of your fuel has been used up, and the food has been eaten during the flight.

a. Is the flight below the maximum allowable weight?

b. Is your center of gravity within legal limits when you take off?

c. The center of gravity changes as fuel burns and as weight (food) is redistributed. Is your center of gravity still within legal limits when you land?

Rena Haldiman
Physics; Astronaut Crew Training Instructor

I was born in Houston, Texas, a city that began as a little cow town. My parents settled down to begin a family in the southwest part of town, when it was still surrounded by fields of grazing cattle, just before it exploded into a big space exploration center for the United States. During President Johnson's time, some land south of Houston owned by Rice University was leased to establish the Johnson Space Center (JSC). It was the newest National Aeronautics and Space Administration site in the country, dedicated to fulfilling President John F. Kennedy's charter of landing a man on the moon before the close of the decade.

Both my parents felt in step with the city's cultural shift towards excellence in science and technology, because each had a background in science. In college, my father had earned a PhD in geology, while my mother had earned a BA in general sciences. They were both lucky enough to live during this dynamic period of U.S. history, influenced, awed, and delighted by the ever-growing importance of scientific knowledge and technological advances.

Motivated and intrigued by their experiences, my parents were also great influences on my and my sisters' appreciation of mathematics and its daily applications in life. Looking back to my childhood, I can remember most vividly the fun, everyday experiences shared with my sisters and parents.

My parents taught us about math simply by sharing and explaining experiences, just for the fun of it. For example, when my mother was baking, she taught us how to follow recipes and measure out ingredients, how to double a recipe or cut it in half for a smaller amount. Counting cupcakes was always fun, and my mom was obviously pleased with our efforts, because she smiled so beautifully all the time.

On the other hand, my father loved after-dinner napkin lessons. He would be so animated about the questions penned out on his napkin that we would laugh and join in, to discover yet another mystery about math, so easily and delightfully explained by our dad. His obvious pleasure in teaching math warmed our interest in the subject.

Both of my parents had learned to play musical instruments during their childhood, so they encouraged us, even providing piano lessons. Once in awhile, we also dabbled with a guitar or flute recorder, trying our hand at reading sheet music written for different instruments. Playing notes for the required duration and emphasis was the key to playing a musical piece. At this young age, we all became familiar with fractions, speaking the wonderful music language of half, quarter, eighth, and sixteenth notes. And the reward was the beautiful music we could summon from the instruments at hand. My parents always expressed great enjoyment when listening to us practice and play our music.

This love of music broadened to include singing, and one of my favorite memories is singing hymns next to my mom in church, hearing her beautiful voice lilt across the words and notes in a way I could only try to copy. I practiced singing alto parts in school choir, learning more about tempo (timing) and becoming more proficient at quickly identifying note types.

These childhood experiences involving math were the foundations for my interest in pursuing further math education. I continued taking math classes throughout high school and college in Baton Rouge, Louisiana, focusing on the science applications, and in particular on physics at Louisiana State University. Physics turns out to be an exploration and explanation of our physical world, based on mathematical descriptions that varied from simple to complex. For me, this is a fascinating and fun way to understand more about my life and the life around me—the dynamics of motion, forces, and pressures—in a variety of applications. Chemistry involves physics at a microscopic and subatomic level, so I took chemistry in college, and worked for two chemistry professors during two semesters. One of them helped me obtain a job offer from the Training Division at Johnson Space Center, near where I was born.

In my first job at JSC , I taught astronauts about the communications systems used by the NASA shuttles to exchange voice and data with the ground controllers at Mission Control Center (MCC). This was not an expected progression for a physics graduate; however, it was technically interesting, and I did use some skills from my college education (e.g., recognizing basic circuitry components on electrical drawings and knowing the logical functions of them). After two years, I was able to move into teaching about shuttle electrical generation and distribution, environmental control, hydraulics, and some mechanical systems, such as payload bay doors and landing gears. This area is closer to my physics interests, and provides endless examples for the study of physics in real world situations.

The mathematical examples I am presenting here represent my current interests in *torque* (forces applied rotationally), and *electricity* (used by the Space Shuttle and in my home).

Gear Ratios in a Shuttle APU System

This example deals with the motion of mechanical gearing—how gears are turned and what turning gears are used for. These questions can be answered when a specific application is analyzed, such as the gears used by the NASA Shuttle's auxiliary power unit (APU). Detailed mathematical descriptions of the system's key parts are necessary, to ensure correct construction and operation of the system. Someone else figured out these details, so I will explain about some basic details, and pose a question about them for the math problem.

Each of the four shuttles in the Shuttle Space Program uses three APUs. An APU's primary job is to drive a hydraulic fluid circulation pump, so the hydraulic fluid and pressure are available at the right places within the shuttle, enabling safe flying and landing. An APU is a turbine generator—a big fan (turbine) with many fan blades close together, with a shaft (like a car wheel axle) attached to it. The shaft is rotated as the fan blades get blown in a circle by the APU exhaust gas. The gas is generated when the APU fuel reacts with the APU's catalyst bed (called an engine). So, chemical energy is converted into mechanical energy, or motion, that is used for turning the shaft. The turbine generator shaft is spun at a speed (revolutions per minute or RPM) of about 74,160 RPM.

At the other end of the turbine generator shaft is a small gear. Gears are toothed wheels that are used for the transmission of motion (rotation) from one part of a machine (the spinning

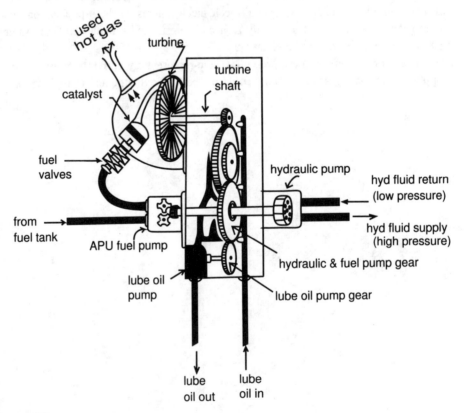

FIGURE 26
Picture of shaft interfacing the three gears

turbine and shaft) to another (the hydraulic fluid circulation pump). The turbine generator shaft gear interfaces with another coupled gear that in turn, interfaces with the hydraulic pump gear (see Figure 26). Since the hydraulic pump gear is bigger than the turbine generator shaft gear, they spin at different speeds or RPMs. The term "gear ratio" mathematically describes the relationship of the different speeds or spins. The gear ratios for the speeds of the turbine generator, hydraulic fluid circulation pump, and the two other necessary pumps used for lubrication (lube) oil circulation and fuel feeding are:

$$\text{Turbine generator shaft to hydraulic pump} = 18.93 : 1$$
$$\text{Turbine generator shaft to lube oil pump} = 6.07 : 1$$
$$\text{Turbine generator shaft to fuel pump} = 18.93 : 1$$

Read this as: "In one minute, the turbine generator shaft completes 18.93 revolutions while the hydraulic pump completes one revolution." Obviously, the hydraulic pump rotates slower relative to the turbine generator shaft. Notice that the fuel pump and hydraulic pump have the same relative gear ratio.

Problem 45. Using these gear ratios, what are the specific speeds, in RPMs, for each of the three pumps driven by the APU turbine generator shaft that operates at 74,160 RPM ?

Electric Power Generation by Shuttle Fuel Cells

Each of the shuttles used by NASA for the Space Shuttle Program has three fuel cells for generating electricity used by the shuttle equipment during a flight in space. The three fuel cells on a shuttle use the chemical reaction of hydrogen and oxygen to produce electricity, water, and heat. The water helps rid the shuttle of heat produced by this reaction and by other equipment operations. The electrical power generated by each fuel cell can be calculated from the operating voltage of the fuel cell, and the load demand (i.e., current demand) of the electrical equipment (i.e., how much equipment is on).

Electrical power has units of "watts," and is calculated by multiplying the operating voltage by the load demand. Each fuel cell is designed to produce a maximum voltage when there is no load on it (no equipment requiring current). As the load demand on a fuel cell increases, the operating voltage decreases. In other words, the voltage and current are indirectly proportional: when one goes up, the other goes down.

(Note: watts = volts × amps, 1 kilowatt = 1000watts)

Problem 46. If fuel cells 1, 2, and 3 are operating at voltages of 30.7 volts, 30.2 volts, and 31.0 volts respectively, and their load demands are 178 amps, 225 amps, and 153 amps respectively, how much electrical power are these three fuel cells collectively producing? And why does fuel cell 2 have the lowest voltage?

Electric Power Usage of a Lamp

Every month the electric bill for my apartment shows the number of kilowatt-hours of electricity that I have used. It is shown in the *kilowatt-hour usage block*. The dollar amount I owe for the

The Light company

ELECTRIC SERVICE BILL

BILL FOR : JAY & RENA HALDIMAN
SERVICE AT: 3807 Friendly Way

Billing Date: 5-22-90	Billing Days: 31	Rate: RESIDENTIAL		
Billing Period	Meter Readings	KWH Multiplier	Kilowatt-Hour Usage	Meter Number
To: 5-19-90 From: 4-18-90	Current: 37928 Previous: 37483	1	KWH: 445	S00602755

Last payment: $22.53   on 5-15-90

— — — BILLING DETAIL — — —   AMOUNT

Residential Service Cost $ 34.20
  Power Cost Recovery Factor - 0.000798 per KWH
Municipal Franchise Fee - 4.2398%   1.45
City Sales Tax - 1%   0.36

| Payment Due 6-12-90 | PAY THIS AMOUNT ===> $ 36.01 |

FIGURE 27
Copy of May 1990 electric bill

month is shown in the *pay this amount* block. *Watts* is the unit for electrical power, and *kilo* means 1000. So *kilowatt-hour* (KWH) refers to the thousands of watts used in an hour.

Problem 47.   On my May 1990 electric bill, usage was 445 KWH and amount due was $36.01. How many dollars were due to my bedside lamp, which has a 75-watt bulb and was on for two hours every night during the 31-day billing period (shown in *billing days* block)?

Elaine Anselm

Business Data Processing

Mathematics was always a fun subject for me. The main reason I liked it so much was because there was always an answer for a given problem. But I was also interested in art, where there was never one right answer.

In 1969, I graduated from an all-girls' parochial high school in Rochester, New York, with a double major in math and art. When I entered college I had a very difficult time deciding between the two subjects. I enjoyed both very much, but the deciding factor was the opportunities for employment—I felt I could make a better living with an educational background in math. Art, in the form of painting and ceramic wheel-throwing, is still a very important part of my life.

I majored in math for two years, then left school to help support my family. With my college math background, I soon found a data processing job as a junior programmer, and continued to attend college in the evenings. I switched my major to computer science at that time, because there were many job opportunities opening in the field and I enjoyed my computer courses and the work I was doing. After ten years of part-time study, I completed the requirements for a Bachelor's Degree in Computer Systems from the Rochester Institute of Technology.

I've been working at Xerox Corporation for about twelve years. For the past year, I've been a training coordinator, developing and teaching technical and business courses for computer professionals at Xerox. I work with instructors from within Xerox, from other companies such as IBM, and from colleges (including the Rochester Institute of Technology), designing courses to be taught at Xerox. Some of the courses we're currently offering include *C++ Programming, Using a Personal Computer, Making Sound Business Decisions,* and *Effective Verbal and Written Communication.*

Prior to working in the Xerox training organization, I was a Technical Information Specialist in the US Marketing Group. In that position, I helped to change the way data processing is done at Xerox. My responsibilities included helping people in business and data processing switch from designing COBOL programs on IBM mainframe computers to designing groups of programs called Business Systems on personal computers.

Up until last year, and for five years before that, I was a database administrator, responsible for understanding and managing the computer storage for most of the information about the customers that buy products and services from Xerox. Some of the database files contain millions of records and must be available to thousands of people, 24 hours a day.

For eleven years before I was hired at Xerox, I was a computer programmer working for a number of different companies in the Rochester area.

These problems relate to the way I used math in my job as a database administrator.

Storing Files on Disk

One way to store business data files is on magnetic computer disks. A magnetic disk is divided into blocks. For this problem assume a disk contains 6,800 blocks. One block can hold from one to 18,400 characters of data. A character is like a single entry on a keyboard (a letter, a number, or a space). Even though a block can contain up to 18,400 characters, information is retrieved and stored (read and written) most efficiently if only 1/4 of the block is used.

A programmer decides the number of records that will be placed in each block for a particular file. The number of characters used by records in a block is called the file block size. To minimize space without concern for how long reading and writing takes, the programmer will fill blocks with as many records as possible (making the block size as close to 18,400 characters as possible). To minimize read-time (read and write the information as efficiently as possible), the programmer will fill only 1/4 of each block with records (making the block size as close to 4,600 characters as possible).

Problem 48.
A. If a file contains records that are each 503 characters long, and there are a total of 115,000 records, how many records should be placed on each disk block to minimize read-and-write-time?

B. When minimizing space for the same file, how many blocks will be used? (Partial records cannot fit in a block except on the last block.)

C. To minimize read-and-write-time without using more than one magnetic disk, how many records can be put in a block?

Storing Information in Memory

Computers store information at addresses in computer memory. A memory address is similar to an address of a house, and the information stored at the address is much like things stored in the house.

The table below represents a piece of computer memory. Addresses are indicated along the top and left margins, and the contents of memory are found to the right and below the bold lines. The left margin represents the ten's, and the top margin represents the one's position of the address. To find the contents of an address, find the intersection between the tens and ones column. What are the contents of memory at address 57? Your answer should be Q.

Problem 49. Once you understand the above example, try to decode the message stored in the computer memory represented below. Start at address 10 and read the contents of three sequential addresses starting at 10. At the first address, find the first character of the decoded message, and the next two addresses (11 and 12) contain the address of the next piece of the message. The hundreds and tens position is at address 11, and the ones position is at address 12. Go to the address indicated and get the second character of the message. Continue following the chain until the final address, FF. Then you will have the entire message.

addresses	0	1	2	3	4	5	6	7	8	9
10	T	1	6		2	6	H	8	5	I
20	4	2			1	9	C	10	7	W
30	3	5		2	9	O	4	5	S	F
40	F		S	8	0	R	5	3	A	
50	W	5	8	K	3	8		Q		7
60	7	L		U	6	6	T	7	4	
70	7	M	10	3	E	9	5	A	1	3
80		8	9	F	F	I	10	0		H
90	9	2	O	5	0	R	3	2		
100	S	2	3	P	6	3		O	7	1

Thinking Outside the Box

At first it may seem that this problem has little to do with data processing, but the point is that not all data-processing problems have straightforward solutions. A great deal of thought and creativity goes into some solutions. Frequently this means approaching a problem from a completely different point of view.

Problem 50. Divide the shape below into four pieces that are exactly similar in size, shape, and area.

FIGURE 28

Smadar Agmon

Software Engineering; Real Estate Investment

At my high school in Israel, everyone had to choose a specialization in biosciences, languages, or mathematics. I chose math because it was my field of interest, as well as a good start for a possible future career. In most of my classes, there were two or three other girls and more than thirty boys. We studied trigonometry, algebra, geometry, and more.

When I was eighteen, I entered the Army for the required two years of service. There I learned more English. I was selected for the officer's course and soon became an officer. This meant I was put in charge of people, some of them older than myself, when I was only eighteen and a half.

After the Army I continued my education. I was interested in both computer science and architecture, so I studied architecture for one semester at the Technion in Haifa, followed by a semester of computer science at the University of Tel Aviv. Although I still enjoyed both disciplines, I chose computer science. In the fall, I returned to Tel Aviv and began to study for a bachelor of science degree. I felt I would rather be a software engineer, because the work gives me lots of problem-solving and clear feedback when I do a good job.

After graduation, I worked for a year in the software division of an insurance company, writing data processing programs in COBOL. I decided to pursue my studies abroad so I could specialize in my field of interest while seeing a foreign country. I entered the University of Toronto for two years, studying for a masters degree in computer science and specializing in databases.

For professional reasons, I wanted to move to "Silicon Valley," the Santa Clara Valley near San Francisco. I sent my resume to several companies there and was invited to interview at some of them. I chose Daisy Systems, staying there for a couple of years. During the second year, my husband quit his job and began a business as a software consultant. While he was getting the business started, I supported him. When my husband's business became more successful, I quit my job at Daisy to work with him. Less than a year later, we hired more people to help us and rented office space to accomodate our new start-up company, Realtime Performance.

Today, Realtime Performance has more than twenty employees. We produce software to control semiconductor manufacturing equipment. Manufacturing these chips is sometimes compared to following a recipe in a cookbook. The people in semiconductor factories use our software to help them control their ovens and ingredients.

At our company, I do various jobs: writing software, managing finances, and managing people. I have also taken several months leave from time to time, so I can travel, spend time with my children, and manage our real estate investments.

The following problems are basic to real estate investment.

Qualifying for Home Mortgage Loans

To qualify for a mortgage loan, a bank adds up all your debts and adds them to the mortgage payments. The total should not exceed 40% of your gross income. (Gross income is your income before paying taxes.)

Example. With a gross income of $40,000 per year, total debts should not exceed $16,000 per year.

    Note. For the purpose of calculating a debt, the bank adds premium and interest amounts. For this exercise we will take only interest payments into consideration.

    These problems start with the easiest and get more complicated as you progress. See how many you can solve.

Problem 51. With $65,000 per year gross income, and a mortgage interest rate of 10% per year, what is the largest mortgage you can qualify for?

Problem 52. With the above income, if you want to buy a house with a 20% down payment and an 80% mortgage loan, what is the highest-priced house you can afford? What will the down payment be?

Problem 53. If you have a car loan with $500 per month payments, and a student loan with $200 per month payments, what is the highest-priced house you can afford?

Problem 54. If in addition to the two existing loans, you also own a rental property with a mortgage loan of $100,000 at 8% yearly interest rate, and a rental income of $1,250 per month, what is the highest-priced house you can afford?

Christine Eckerle
Quality Engineering

I've always loved reading mystery stories, and my job involves solving problems, just like they do in the stories. I look at all of the things that might be causing the problem, and design experiments to find out what the main causes are, so a solution can be found. Sometimes those solutions are "people processes," or changes to the way something is manufactured. To help find the solutions, I use mathematics and logic tools such as cause-and-effect diagrams, histograms, averages, storyboarding, and many other techniques. I work with people and computers, doing a lot of research to find the answers. When I solve a difficult problem, I feel a great sense of satisfaction—and pure joy.

In high school, I didn't know what I wanted to do, so I tried to take classes that would prepare me for various careers. Taking three years of mathematics gave me more options when I went to college. At Saginaw Valley State University, I earned a BS in Mathematics and an MBA. Then I taught for 12 years in elementary and high schools, before I started working for Delta College as a quality engineer in both the manufacturing and service industries. Now I'm an educator/trainer for Corporate Services, meeting the needs of the surrounding businesses. Recently, I've been on loan to Saginaw Division, General Motors Corp. as a consulting Quality Engineer, Quality Auditor, and Quality Assurance trainer.

As a child, I wanted to solve mysteries when I grew up, and I have achieved my dream. I didn't know about quality engineering when I was young, but when I decided that's what I wanted to do, my math education gave me all the skills I needed to enter this profession. Without a background in math, I wouldn't have this job—one I really enjoy and that is so much fun.

Machine Capability

When a new production machine is purchased, the quality engineer wants to know whether it can make the parts needed by the customer. The customer provides a blueprint describing various dimensions of the part. Each dimension has a nominal value, or target, and a tolerance that is the allowable deviation from that target.

The machine operator tries to make the parts right on-target, but due to variability in the machine and the process, it isn't always possible to make the parts exactly the way the blueprint says they should be. The tolerance tells the engineer how far off-target the part can be and still be used by the customer. As the process evolves, the quality engineer tries to make continuous improvements by reducing the variability. But the baseline variability measurement must be established first.

So, the quality engineer performs a machine capability by measuring parts that have been controlled for outside sources of variability. Using the same operator and the same material (i.e., the same heat of steel), the engineer selects parts for measuring over a short timeframe, isolating only the variability produced by the machine. A machine capability involves taking a sample of parts over time, and then looking at the measurements statistically. The result is compared to the allowable tolerance described by the blueprint. Then the machine capability is listed by some of the various measures of capability such as CP, CR, or CPK.

Definitions:

$$s = \text{standard deviation of a sample}$$

$$\bar{x} = \text{average of the sample}$$

$$\bar{\bar{x}} = \text{grand average (the average of averages)}$$

$$\bar{r} = \text{average of the ranges}$$

$$\text{USL} = \text{upper spec limit (from blueprint)}$$

$$\text{LSL} = \text{lower spec limit (from blueprint)}$$

$$A_2 = \text{factor from chart for control limits on averages}$$

$$D_3 = \text{factor from chart for control limits on the range}$$

$$D_4 = \text{factor from chart for control limits on the range}$$

Note: A_2, D_3, and D_4 are based on sample size.

The CP is the capability of the process. It is measured by dividing six standard deviations ($6s$) into the tolerance. The standard deviation is found by statistically analyzing the measurements of the parts for that dimension. The tolerance is found on the blueprint.

$$\text{CP} = \text{tolerance}/6s$$

The CR is the capability ratio of the process. It is the reciprocal of the CP and is found by dividing the six standard deviations ($6s$) by the tolerance. This number is then changed to a percentage.

$$\text{CR} = 6s/\text{tolerance}$$

The CPK is a measure of where the center of the process is in relation to the tolerance. The letters come from the Japanese word for target. The CPK is found by using the average ($\bar{x}$) of the process and the standard deviation (s).

$$\text{CPK} = \min\left[(\text{USL} - \bar{x})/3s, (\bar{x} - \text{LSL})/3s\right]$$

These measures may not be valid if the process is not stable. So the quality engineer must also check the stability of the process by determining if the process is in statistical control. This is done by taking those same measurements, putting them on a control chart, determining the control limits, and then interpreting the data based on the chart. If the process is stable or predictable, then the points will all be within the control limits and won't show any grouping within the limits.

The formulas for control limits are as follows.

For Center:

$$\text{UCL} = \bar{\bar{x}} + A_2 \bar{r}$$

$$\text{LCL} = \bar{\bar{x}} - A_2 \bar{r}$$

For Range:

$$\text{UCL} = D_4 \bar{r}$$

$$\text{LCL} = D_3 \bar{r}$$

For a machine to be capable the guidelines are:

$$\text{CP} > 1.33$$

$$\text{CR} < 75\%$$

$$\text{CPK} > 1.33$$

Problem 55. A quality engineer performed a machine capability and measured 125 parts. After determining the average and standard deviation, the capability ratio (CR), the CP and CPK must also be determined to satisfy the customer's requirements. Given an average ($\bar{x}$) = 15.3 mm and a standard deviation (s) of .25mm, find the CR, CP, and CPK. In this case, $\bar{x}$ and $\bar{\bar{x}}$ are the same. The tolerance is 15 mm $\pm$ 1 mm. The quality engineer must also determine whether the process is stable or not by analyzing the control chart. The $\bar{r}$ = .15 mm, and the subgroup size is 3. Determine the control limits for the chart. For subgroup size of 3:

$$A_2 = 1.023$$

$$D_3 = 0$$

$$D_4 = 2.547$$

Average time

Janine wanted to find out how much time she spent every night, doing her homework. She thought she had too much homework, but her mother felt just the opposite. Janine and her mother decided to keep track of the time spent, so they could find out who was right.

First, they had to define just what working on homework meant. Janine's mother said that if Janine was watching TV, or talking to her friends on the phone while doing homework, that time shouldn't count. They discussed this and came up with an operational definition, so Janine knew exactly when she was really doing homework.

Every day, Janine kept track of the minutes when she did homework. They were as follows:

Week 1			Week 2		
	Monday	65 min.		Monday	70 min.
	Tuesday	35 min.		Tuesday	25 min.
	Wednesday	40 min.		Wednesday	55 min.
	Thursday	30 min.		Thursday	40 min.
	Sunday	50 min.		Sunday	60 min.

Problem 56. What was the average length of time that Janine spent on her homework?

Quality Costs

Quality Engineers must talk the language of management—money. So they describe their actions using four different categories, and then use dollar amounts to compare their performance to the budget set by the company. These four categories of quality costs are: prevention, appraisal, internal failure, and external failure. These categories are divided into activities, so performance can be tracked to show whether those areas are improving from month to month.

Prevention activities are designed to prevent the product or service from being made incorrectly. These activities include training, quality planning, and supplier quality evaluation. Appraisal activities are inspection functions that help decide if the product or service meets the customer's requirements. Failure costs are broken into two components, depending on whether the company or the customer discover something wrong with the product or service.

Below is a listing of the quality costs for the Apez Company for the month of October:

Prevention:

training	150 man-hours	@ $30. per hr.
design review	40 man-hours	@ $35. per hr.
audits	20 man-hours	@ $25. per hr.

Appraisal:

process control	250 man-hours	@ $25. per hr.
incoming inspection	50 man-hours	@ $25. per hr.

Failure:

scrap	$55,000.
rework	$70,000.
warranty	$35,000.

Problem 57. What was the cost of quality for the Apez Company for the month? Which category of quality costs were the largest?

Sally Irene Lipsey
Health Science

At Hunter College High School, I loved math and took almost all the courses that were offered, including trigonometry and solid geometry, but not calculus. I went on to Hunter College where I majored in math with an education minor. I became an assistant in the math department at the University of Wisconsin, where I also earned an MA in math.

Newly married, I next went to the Columbia University School of Pure Science for a PhD. I completed all the courses for the degree, but acquired a family instead—three daughters in four years! While my daughters were young, I worked as an assistant professor of mathematics at Bronx Community College. It was the early 1960's, when a teaching method known as "Programmed Instruction" was very popular, almost a fad. It consisted of a teacher-student dialogue, written in the form of questions and answers.

One day the New York State Department of Education contacted our college, asking for urgent help teaching math to nurses. Although they were otherwise good nurses, many of them had been so badly trained in mathematics, or had so little faith in their ability to do math, that they were actually causing health problems. Poor math skills were causing them to make errors in documentation and dosage calculations.

Using Programmed Instruction to teach math to nurses seemed like an interesting doctoral project, so I returned to Columbia's Teachers College and completed an EdD. This resulted in the publication (by Wiley) of *Mathematics for Nursing Science* in 1965, with a second edition in 1977. Although Programmed Instruction is no longer an important style for textbook-writing, it still plays a role in computer-assisted instruction.

After 1965, I taught at Brooklyn College, using computer-assisted instruction as a tool to help students in every field brush up their math skills. I was also associated with the School of Education in the preparation of teachers. In 1985, I chose early retirement from my position as Associate Professor of Mathematics at Brooklyn College. I continued to write for nurses, but no longer in the Programmed Instruction style.

Recently I had the pleasure of writing a book with Donna Ignatavicius, a nurse. Our book, *Math for Nurses: A Problem-solving Approach,* was published by Saunders in 1993. With Donna's help, I discovered more about the medical side of nursing than I had ever learned through my own research.

The problems in this latest book are directly related to the way nurses really use math in their daily work.

The following problems are all taken from the health sciences, where simple math is frequently a matter of life or death.

Weight-based Dosage

Problem 58. As preparation for surgery, a patient is scheduled to receive an intramuscular injection of scopolamine, 0.005 mg/kg. If the patient weighs 156 lb, how many mg of scopolamine should she receive?

Critical Care Beds

Problem 59. A hospital in Los Angeles has 360 beds. Thirty of them are for critical care. At a smaller hospital in the city, there are 164 beds, but only 14 are for critical care. Which hospital has a greater percentage of critical care beds?

Cholesterol Analysis

Problem 60. The measurement (in milligrams per deciliter) of low-density lipoprotein cholesterol (LDL) is important in health assessment and management. Guidelines are often given as follows:

$$LDL < 130, \quad \text{desirable};$$

$$130 \leq LDL < 160, \quad \text{borderline-high};$$

$$LDL \geq 160, \quad \text{high-risk}.$$

To estimate low-density lipoprotein cholesterol, first measure total cholesterol (C), triglycerides (T), and high-density lipoprotein cholesterol (H). Then, if T < 400, use the following formula:

$$LDL = C - H - T/5,$$

where T < 400, and all measurements indicate the number of milligrams per deciliter.

A lipoprotein analysis done for a certain woman showed C = 230, H = 50, and T = 80. Which category of the guidelines does she fit into?

Janean D. Bowen
Nursing Education

I am currently in my sixth year of teaching in the Practical Nursing Program at the Northeast Kansas Area Vocational Technical School. One of my courses is a basic math refresher to prepare students for pharmacology classes. The majority of my students are women who, at some time in their lives, were told they didn't need to learn math. Many don't understand why nurses need to know how to use fractions and proportions. When I began my nursing career, I didn't know why, either.

In high school, I had no clear career plan in mind. I took four years of foreign language and thought I'd probably pursue that area in college. However, because I wanted a broad preparation for college, I also took three years of math: algebra, geometry, and advanced algebra, including some trigonometry. Math did not come easily to me. I spent a lot of time at the kitchen table with my father. He knew little about math himself, but using logic and experience, he usually found a solution I could understand.

My high school had no college counselor, so I chose a small women's college and enrolled in what I thought was a liberal arts track. Six weeks into classes I found that I was classified as a pre-clinic in a three-year nursing program! I decided to complete the semester so I wouldn't lose my credit hours. But as I continued, I found that I enjoyed science classes the most. And I liked the idea of helping people as a nurse.

To become a registered nurse, candidates must successfully complete a nursing program and pass the Nursing Board Examination. Although nursing schools now offer three types of nursing programs—a two-year (Associate Degree) program, a three-year (Diploma) program, and a four-year (Baccalaureate Degree) program—the three-year programs are being phased out in favor of the academic programs that require more math.

I chose the three-year (Diploma) program, with no formal math classes. In a brief segment of the pharmacology course, we learned conversion equivalents for metric, apothecary, and household measurements. Students also had to calculate a few drug dosages using proportions.

My first twenty years as a nurse were spent in hospitals providing direct patient care. I tell my students that not a day went by that I didn't use math in some way. Some uses are obvious—when we calculate a drug dose. Some uses are only for very specialized areas of the hospital, like the intensive care unit, where we calculate the amount of blood flow through the heart in one minute.

Ten years after my graduation, I began courses to complete requirements for a Bachelor's Degree in Nursing. I took *Basic Math* (as a refresher), *Calculus I, Statistics,* and a beginner's course in computer programming. In July, 1993, I completed requirements for a Master's Degree, including analyzing statistics from a research project I conducted.

The problems that follow are every-day situations that require nurses to do some calculating.

Intravenous Therapy

Equivalents

METRIC	APOTHECARY	HOUSEHOLD
60 mg	1 grain	—
100 mg	1 1/2 grains	—
4 ml	1 dram	1 teaspoon
30 ml	1 ounce	1 ounce
500 ml	1 pint	1 pint
1000 ml	1 quart	1 quart
1 kg	—	2.2 pounds

When patients are admitted to the hospital, they frequently need fluid replacement, usually given into the veins. This intravenous therapy uses tubing and a drip chamber to control the flow of fluid into the patient's arm.

The manufacturer of the tubing has measured the drop mechanism (diameter of the dropper) and found it to yield 1 ml (milliliter) for every 15 drops of fluid. This information is printed on the tubing package. When ordering fluids, the doctor will order either the total amount of fluids to be given in a specific time frame or the number of ml to be given in one hour. The nurse must be able to calculate the number of drops per minute to equal the amount of fluid ordered for the patient.

Problem 61. The order for Bill Jones reads: 125 ml per hour. To carry out this order correctly, the nurse must regulate the drip rate to how many drops per minute?

Problem 62. The order for Mary Brown reads: 1 liter of fluid every ten hours. How is the drip rate calculated for one minute?

Dosage Calculations and Conversions

Medications are frequently mixed or calculated based specifically on each patient's weight, usually measured in kilograms. This technique is especially good for children, because they can be overdosed so easily. All four-year olds are not the same size!

When working out any problem, all parameters must be in the same number system. Unfortunately, the medical community uses apothecary measurements (ounces and grains), household measurements (teaspoons and drops), and metric measurements (milligrams for weight and milliliters for volume). These measurements must be converted to the same system to calculate the dose for each patient.

Problem 63. Paul needs an antibiotic for tonsilitis. Amoxicillin is ordered. Paul weighs 13 kg. The medication comes as a liquid containing 250 mg per 5 ml. The package insert states the safe dose is 20 mg per kg. What is the safe dose for Paul? How much medication would be given?

Problem 64. Mandy, age four, needs digoxin, a powerful medication used to regulate and strengthen the heart beat. She weighs 40 pounds. The dosage information in a drug reference states that the safe dosage for children is 40 mcg (micrograms) per kilogram per day. This medication is supplied as an oral liquid containing 50 mcg per ml. What is the safe dose for Mandy? How much medication would be given in 24 hours?

Problem 65. Valium for injection is supplied as 10 mg per 2 ml. The doctor orders 3 mg. What volume will be given?

Problem 66. Morphine sulfate for injection is supplied as 10 mg per ml. The doctor orders $\frac{1}{10}$ gr. (grain). What volume will be given?

Problem 67. Thyroid hormone replacement is dispensed in micrograms, milligrams, and grains. Being able to interchange these forms means the nurse can save time and reduce errors.

Minnie is 85 years old. Her last thyroid test revealed that she needed a change in her medication dosage. She had previously taken 0.5 grains daily. Her new prescription is for 15 mg daily. Did her dose increase or decrease?

Problem 68. Sam is low in thyroid hormone and will begin taking this drug daily. His prescription reads 0.3 mg per day; the bottle label reads 300 mcg. Has a mistake been made by the pharmacy?

Amy C. R. Gerson

Electrical Engineering; Space Systems

I still remember how my first grade teacher, Mrs. Perkins, let us watch the Apollo rocket launches on television. The powerful technology inspired me to be a good student, so I could be a part of the space program. I enjoyed school and got a lot of encouragement from my family. My mom is an entomologist (a scientist who studies insects), and she has also been an inspiration. I still enjoy visiting her lab and peering through the microscope.

In high school I studied algebra, geometry, trigonometry, and calculus. As I progressed, I gradually decided that even though the language arts, social studies, and French courses I took were easier, I still wanted a career in math or science. Those fields offer challenging jobs that pay well, and I still dreamed of being involved in the U.S. space program. I was also attracted to the mystique surrounding math and science. They were subjects with a certain prestige, ones that not just anyone could master.

Eventually, I realized that math and science weren't that mysterious, and I could successfully pursue them. At the University of Washington, I studied electrical engineering, because I thought it was more versatile than aeronautical/astronautical engineering. When I graduated in 1984, I received eight job offers. I selected the Boeing Company because it gave me a chance to help develop the next generation of space solar power technology. Since then, I've enjoyed the excitement of working on several space projects, including the Space Station, the Lunar Resource Mapper, and the Mars Environmental Survey Network. I've also written four research papers and hold a patent for a high-voltage space solar array design.

After I'd been working at Boeing for awhile, I began to miss the academic environment. Since I also wanted to learn a new subject, I pursued a master's degree in electrical engineering part-time. Boeing paid my tuition.

As part of my graduate work, I took courses in automatic control and completed a research project on chaos theory. For a design project, I joined a team of students who built a successful microgravity experiment that flew on the Space Shuttle in the fall of 1992.

Recently, Boeing sent me to a summer session of International Space University, where I studied ten space-related subjects with 100 people from 28 countries. Students from all over the world cooperated to complete a design study for an international astronomical observatory on the far side of the moon.

There haven't been many women in my career path during my college years or at work, but I've made some very close, valuable friendships with many of the women I did encounter. We share support, understanding, and a sense of humor about our situation. Our male colleagues are lucky to know us—we brighten up the office. In my personal life, I am happily married to a mathematician, and we have an Alaskan Malamute dog for a pet.

Satellite Electrical Power System

Consider the following electrical power system for a satellite in low earth orbit. While the satellite's solar array is illuminated by the sun, the array generates power and recharges the spacecraft's batteries. These batteries are needed when the spacecraft demands more power than the solar array can provide, for special maneuvers and during eclipse periods.

At the beginning of a 90-minute orbit, the spacecraft performs a 15-minute maneuver that requires more current than the maximum solar array capability, causing the batteries to discharge to make up the deficit. The maneuver causes a maximum battery discharge of ten amperes. During the maneuver, the battery current varies sinusoidally with period 30 minutes and phase shift zero. From 15 minutes to 60 minutes, the solar array is capable of supplying the total spacecraft load, and the batteries do not discharge. During the last 30 minutes of the 90-minute orbit, the spacecraft is in the shadow of the earth, so the batteries must supply the total spacecraft load of 30 amperes.

To avoid damage to the batteries, spacecraft operations are planned in advance and checked against battery discharge constraints. Cumulative battery discharge is calculated for each orbit for comparison against a maximum allowable limit.

Problems 69. Part a.   Battery discharge in amperes is a function of time. Plot the function, giving battery discharge in amperes as a function of time for the 90-minute orbit described above.

Part b.   Write an equation for the battery discharge function.

Part c.   Calculate the cumulative battery discharge (in units of ampere-hours) for the 90-minute orbit described above.

Part d.   Determine whether the cumulative battery discharge exceeds 40% of battery capacity, assuming battery capacity is 50 ampere-hours. If the cumulative battery discharge does not exceed 40% of battery capacity, the batteries will not be damaged by the proposed operation and the mission may proceed as planned.

Note.   Part c requires calculus to find the area under the curve. How might you approximate that area without using calculus?

Marilyn K. Halpin
Oil and Gas Accounting

As a junior and senior high school student, I did well in all my classes, but math class was always the easiest. In junior high, my math teachers were two stern, former military men. They drilled us with pages and pages of problems that were like puzzles to me. Finally Elizabeth Mann, my trigonometry teacher and coincidentally my Student Council sponsor, led me toward a career in mathematics. To this day, trigonometry is my favorite course.

With no detours, I became a mathematics major at the University of Texas in Austin. Between earning a BA (1969) and an MA (1971), I taught remedial math at my former high school. My students had always failed in math, so their perspective broadened mine toward both teaching and math. During graduate school, I was a teaching assistant in two courses: math for business majors, and calculus.

Because I was so interested in education, my first job after college was as a consultant for the Texas Education Agency (the state's Department of Education). I conducted inservice training workshops in drug education and crime prevention for teachers. The goal of the program was to show teachers how to use hands-on activities that would involve their students in decision-making and communication. For example, in a US history course, students used a self-assessment instrument to determine their personal values about gun laws. Forming into small groups, they reached a consensus about what type of gun laws they would introduce if they were state legislators. Participants in the training completed attitude surveys I had developed to evaluate the program and measure the workshops' effectiveness.

I was initially reluctant to join the computer world. In those days before personal computers and interactive programming, I was turned off by computer cards, bits, and bytes. However, once I accepted an entry-level position as a programmer in 1978 and sat down at my own terminal, I was hooked. For several years, I programmed software to produce tax returns used by accounting firms, and also managed programmers and accountants. Then I went to work for a banking software company as an instructor for bank programming personnel. This job combined teaching and programming, suiting me perfectly.

In 1987, I began working for my husband, a petroleum geologist. As his administrative assistant, my primary responsibility is to maintain the books, so I have evolved into an oil-and-gas accountant. I record revenue and expenses each month utilizing specialized accounting software for the oil-and-gas business. Since teaching is still a strong interest, I tutor high school geometry and algebra two mornings a week.

Gas Settlement Problem

During the early 1980's, natural gas was in short supply. Independent producers were able to secure long-term contracts with gas purchasers, at high prices. The gas contracts specified both a contract price and an amount of gas to be purchased. As the decade came to an end, gas prices fell dramatically. In some cases, prices dropped to one-fourth of their former levels.

Purchasers did not want to buy gas at the higher price, so they offered "take or pay" settlements to producers. These settlements are cash payments for gas that was supposed to be delivered but was not. Producers receive the cash, and the money is recouped from future gas sales.

For example, Bob Thornbury is an independent producer who had a contract with ARKTX. ARKTX failed to buy the gas specified in the contract. After negotiations, Mr. Thornbury agrees to receive $30,673.48 as a settlement from ARKTX. The money will be recouped at a rate of 75% of future gas purchases.

Mr. Thornbury also owns part of a natural gas well. His interest in the well is .095756. The following chart lists the gross gas production in mcf (thousand cubic feet) from November, 1989 through October, 1990:

11-89	1806 mcf	12-89	3354 mcf
1-90	3112 mcf	2-90	2777 mcf
3-90	2858 mcf	4-90	2629 mcf
5-90	2769 mcf	6-90	2567 mcf
7-90	2578 mcf	8-90	2590 mcf
9-90	2508 mcf	10-90	2494 mcf

First, compute the mcf Mr. Thornbury is entitled to each month. Since he owns only part of the well, he is entitled to part of its gross gas production. Another factor determining Mr. Thornbury's proper monthly allocation is the BTU (British Thermal Unit) rate. This rate is a heat-equivalent factor. Before multiplying the net production by the price, the net production must first be multiplied by the BTU rate, in effect adjusting for the "energy value," or quality, of the gas the well produced.

For Mr. Thornbury, the BTU rate is 1.13342 for each month except October, 1990. For October, 1990, the BTU rate is 1.12632. Allocation is mmBTU (million BTU). The price of gas for each month is $3.00/mmBTU.

Problem 70. Part a.   What is the total earned income for Mr. Thornbury during this period?

   Part b. What is the amount recouped by ARKTX?

   Part c. How much has Mr. Thornbury actually received?

   Part d. How much does Mr. Thornbury still owe ARKTX to repay his settlement?

Martha Leva

Business Administration Higher Education

Mathematics always made me feel successful. My grade school was very small, less than 250 students, with more girls than boys in all of my classes. All of my teachers were women, and girls were always the top performers in math.

As time went on, I attended an all-girl high school and took all the math courses that were offered to college-bound students, including calculus during my senior year. All my math teachers were very serious and dedicated women. I specifically remember my geometry and algebra teachers, who made the material very easy to understand.

Following high school, I attended a co-ed liberal arts college. In my freshman and sophomore years, I fell in love with art, music, and literature, and filled my schedule with these new interests. But in my junior year, I gravitated back to math, taking a calculus course. For the first time in my life, the teacher at the front of the class was a man. Also for the first time, I was outnumbered by the men in the class. But that didn't matter. I still enjoyed the work and was successful at it.

My junior and senior years as a psychology major required specific courses including statistics, research design, and logic. Newly-offered computer science courses attracted me, so I enrolled in an introductory course that focused on machine language. The professor held a doctorate in computer science and was my first non-liberal arts woman professor. She was wonderful! She opened a whole new world, encouraging me to study FORTRAN and COBOL during my last semester as an undergraduate.

After graduation, I worked in the social service sector, developing educational programs for the public aimed at preventing mental illness. Along with the rest of my team, I spoke to high school students and community groups about developing a healthy outlook on life and preventing psychological problems. These programs were funded by the state and federal governments, and required budget proposals, record-keeping, and financial reporting. Working in this field taught me the importance of sound business practices, and pointed out my lack of business expertise.

Now my career changed direction. I returned to school in pursuit of a Master's degree in Business Administration at Temple University in Philadelphia. As a graduate student, I studied two additional statistics courses; several courses in accounting and finance; and operations research (applying mathematics to business problems). To afford full-time graduate study, I applied for and was granted a graduate assistantship. This covered my tuition and paid a small stipend. In exchange, I assisted an economist in his research on the electric utilities industry. My computer training proved invaluable because I had to retrieve data from a computerized database to analyze factors that influence customers' electric bills.

After graduation, I began studying for my PhD while teaching college-level business courses at Penn State University. I married and became a mother. Now I teach business courses at the Ogontz Campus of Penn State University, near Philadelphia. I enjoy teaching a variety of courses—all with math overtones and undertones: accounting, corporate finance, personal finance, small business problems, quantitative business analysis, and statistics.

My two daughters pleasantly divert me from continuing my studies. I happily try to teach them what I know about computers on their own level. At the ages of four and seven years old, both of them can use a mouse, open and close windows, and even change drives on our home computer!

Formatted Stock Prices

While gathering information for a research project about United States' corporations, I discovered that the stock prices were recorded in the computer in a shortened form, making it impossible to use them in arithmetic calculations.

Stocks are traded in eighths of a dollar. For example, a price of 12 7/8 means that the stock is valued at twelve dollars and eighty-seven and one-half cents ($12.87 1/2). In the computer record, the same price would be recorded at 12.7.

Note that this is not a mistake. The prices are recorded this way to save computer memory. In decimal form, the fraction 12 7/8 is 12.875. Using the formatted stock price "12.7" to represent 12.875 means that the computer must store only one digit after the decimal instead of three. Yet these formatted stock prices are useless for arithmetic calculations—for example, summing up the stock price each day for a month in order to calculate the average price—because the number after the decimal place represents eighths instead of tenths.

Problem 71.   How can you use a computer language to change the data to a dollar-and-cents-figure that can be used in arithmetic calculations?

(If you don't know a computer language, just make one up. This is what software engineers do when they write a program. First they write it in "pseudo-code," their own shorthand language for specifying algorithms. Then they re-write it using a computer language.)

Using your solution, change these formatted stock prices into real decimal numbers: 29.0, 30.1, 10.2, 15.3, 112.4, 20.5, 50.6, 33.7.

Why would a formatted stock price of 11.8 or 15.9 be an error?

Caroline P. Nguyen
Aerospace Engineering

When I came to the United States fifteen years ago, I knew very little English. In fact, the only complete sentence I could say was "my name is . . . ." Due to political reasons, I was forced to leave my country, Vietnam, before I completed junior high school. However, after enrolling in high school in America, I realized that it would take a miracle for me to earn a high school diploma. The difficulties were compounded by the fact that I didn't know the language and I had to assimilate a completely new culture in a fast pace. Because of the English handicap, math was the only thing that kept me in school, since it was taught in the universal language of numbers. Even though I struggled with word problems, I continued taking up to three math classes per semester to compensate for other classes. My first biology class was an unforgettabe experience; most of my biology reports were prepared partly in English and partly in my language. I waved my hands to make conversation with other students during laboratory hours.

After graduation from high school as an honor student, I went to college to fullfill my parents' wishes. The extra math I had taken in high school allowed me to skip three semeters of college-level calculus. However, language difficulties continued to exert greater pressure on me as I advanced toward a college degree. To cope, I recorded most of the lectures, replaying them until I could understand them.

I graduated from U.C. Berkeley with a BS degree in Chemical Engineering, and decided to continue my education so I could fullfill my dream of becoming a technical consultant for development firms. I entered the masters program at the University of Washington at Seattle, earning an MS degree in Chemical Engineering. To support myself in college, I worked as a research assistance, earning a second background in Electrical Engineeering.

Now I am an Aerospace Engineer. My speciality is energy storage systems for spacecraft and missiles. I've also worked on hazardous-waste reduction and pollution-prevention projects, replacing certain environmentally-threatening chemicals currently used in the aerospace industry. My achievements include a Hubble Space Telescope Achievement award, an inclusion in the 18th Edition of *American Men & Women of Science,* and a National ART award.

My English is much better than it was 15 years ago. I really enjoy learning and continue to take classes, which have earned me certificates in areas such as statistic and system engineering. With all the technology in today's society, like Lewis Carroll (a famous writer) said, "It takes all the running (learning) you can do, to keep in the same place."

Sizing a Spacecraft Energy Storage System

Space systems have played an important role in today's technology. They support worldwide communication, provide weather information, and enhance operations through all levels of military and strategic defense. If you are curious about how these spacecraft power themselves in space, the description below will give you a general concept of the spacecraft's energy sources.

Spacecraft convert the solar (sun) energy, readily available in space, into electrical energy. This electrical energy is generated by solar arrays, sometimes known as the "wings" of the spacecraft. These solar arrays consist of hundreds of solar cells that make electricity when the sun rays pass through the cells. However, when the spacecraft enter the dark side, where the sun is blocked by the earth, they need another source of power. So they carry their own energy storage system, also known as the "battery" system, that is similar in function to a car battery system. The battery is a device that stores chemical energy that can be converted to electrical energy when needed.

There are two major types of energy storage systems: primary (one-time use only), and secondary (rechargeable). If the mission duration is over two years, the rechargeable system will most likely be chosen. And since most spacecraft are very expensive (costing millions of dollars apiece), the mission duration is usually designed to be as long as possible. Currently, there are two established rechargeable systems for spacecraft: nickel cadmium (NiCd) and nickel hydrogen (NiH$_2$). The NiCd system is losing popularity because there is more demand for higher-power missions. Also, it is expensive to manufacture cadmium, a chemical that can potentially contaminate the environment. Other systems, such as sodium sulfur and lithium polymer, are also being developed for future space applications.

Before discussing the energy systems, the following terms should be defined:

Load. the amount of power needed to operate a spacecraft once it is in space (orbit). The unit of load is watts (w).

Capacity. the total amount of energy that can be withdrawn from each battery at a specific voltage. The unit of capacity is ampere-hour (AH). This term is normally seen on car batteries. If the amount of current (ampere) withdraw is high, then it will take less time to deplete the total battery potential.

Percent Depth-of-Discharge (%DOD). the ratio of the amount of capacity desired to the amount of capacity available within the battery. Higher %DOD can result in shorter battery service life. On the other hand, fewer batteries will be needed, adding up to lower total cost and weight. The selection of %DOD depends on three parameters: mission duration, payload requirement, and the chemical type of the energy storage system (nickel hydrogen, nickel cadmium). A good selection of %DOD can be derived through an iteration process, to optimize the power output for spacecraft. The rule of thumb for LEO orbit is ten to 20 percent; MEO orbit is 20 to 50 percent; and GEO orbit is 50 to 75 percent.

Battery Efficiency. the ratio of the output current (or discharging) to the input current (or charging). This number varies with the battery chemistry and the design. The efficiency is also decreased with service time. Similar to %DOD, higher battery efficiency also translates to lower overall cost and weight.

In addition, there are three major types of orbits for spacecraft: Low Earth Orbit (LEO), Mid-attitude Earth Orbit (MEO), and Geosynchronous (GEO).

A general concept of each is described below:

LEO.   It takes a total of 95 minutes for a spacecraft to orbit around the earth in this orbit. There are only 60 minutes of sun and 35 minutes of darkness. In 24 hours, the spacecraft will complete 16 orbits. The distance from this orbit to the earth is approximately 600 nm (nautical miles).

MEO.   It takes a total of six hours for a spacecraft to orbit around the earth in this orbit. There are 5.25 hours of sun and 0.75 hours of darkness. In 24 hours, the spacecraft will complete four orbits. The distance from this orbit to the earth is approximately 5600 nm.

GEO.   It takes a total of 24 hours for a spacecraft to orbit around the earth, or to complete only one cycle in this orbit. There are 22.8 hours of sun and 1.2 hours of darkness. The distance from this orbit to the earth is approximately 22,500 nm.

In addition, hybrid orbits such as LEO-MEO or MEO-GEO are also possible.

Sample Problem.   Since the energy storage system interfaces with other components within the spacecraft, the parameters described below are needed to perform battery sizing:

1. Type of orbit and the orbit time
2. Size of the spacecraft (payload) and the voltage requirement for the operation of the payload (BUS)
3. Percent Depth-of-Discharge (% DOD). In this sample problem, 15% DOD is selected for a nickel hydrogen (NiH_2) battery system to ensure a mission duration of five years. (This implies that only 15% of the battery capacity is used during discharge or dark period.)
4. Battery efficiency

For the purpose of this discussion, translate the above conditions to numbers so a calculation can be performed.

Given Conditions:

LEO orbit	60 minutes of sun and 35 minutes of darkness
Spacecraft size	4000 watts payload at 28V operation
Battery system	Selected NiH_2 system with 15% DOD for five years operation
Battery efficiency	For NiH_2, roughly 80%, or 20% less than the theoretical value.

Solution of Sample Problem.   Since the battery efficiency is only 80 percent of the theoretical value, the overall battery system must be sized with 20 percent margin to sufficiently support the 4000 watts payload. (With 80% efficiency, the battery can only deliver 3200 watts.)

Actual spacecraft payload required = 4000 watts/0.8 = 5000 watts.

On the other hand, the spacecraft needs power only during the dark period of 35 minutes.

Spacecraft payload required/LEO orbit = 5000 watts × (35 min/60 min per hour)

$$= 2917 \text{ WH}.$$

For a 28V BUS spacecraft, the payload (WH) required can be converted to the capacity (AH) required, a more universal unit for the battery-sizing process.

Watts = Volts × Amps,   Watts-Hours = Volts × Amps-Hours.

Spacecraft capacity required/LEO orbit = 2917 WH/28V BUS = 104AH.

However, we have selected the NiH_2 battery system with only 15% DOD. Therefore, the total battery capacity required for the spacecraft must be recalculated.

Battery capacity required = 104 AH/0.15 = 695 AH.

Finally, the number of batteries needed can be calculated by selecting a battery brand from various battery suppliers. For the purpose of this discussion, let us assume a 90AH NiH_2 battery size has been selected.

Total number of batteries needed = 695 AH/90 AH = 8 batteries.

The total weight and size of the energy storage system can be calculated using the weight and size of each battery, multiplying each by a factor of eight. The Hubble Space Telescope spacecraft consists of six 90AH NiH_2 batteries with a total weight of 920 lbs. Without its protection cover, each battery weighs 127 lbs.

Problem 72. The process of sizing the energy storage system for a spacecraft is slightly more demanding in the real world, but the basic concept is the same.

Since the previous problem involved sizing the energy storage system for LEO orbit, it would be interesting to size an energy storage system for the other two orbital types, MEO and GEO.

Given Conditions:

MEO orbit	5.25 hours of sun and 0.75 hours of darkness
GEO orbit	22.8 hours of sun and 1.2 hours of darkness
Spacecraft size	4000 watts payload at 28V BUS
Battery system	NiH_2 with 40% DOD for MEO orbit
	NiH_2 with 65% DOD for GEO orbit
	NiH_2 battery efficiency is 80%

Follow these steps, as in the Sample Problem, to figure out how many batteries are needed for a spacecraft in MEO and GEO orbits:

1. Calculate the actual spacecraft power required.

2. Calculate the actual spacecraft power required for MEO and GEO orbit.

3. Calculate the actual spacecraft capacity required for MEO and GEO orbit.

4. Calculate the actual battery capacity required for MEO and GEO orbit.

5. Calculate the total number of battery needed per MEO and GEO orbit.

In addition to the above, you may want to ask yourself the following questions:

How many times does a spacecraft orbit around the earth in one year at LEO orbit? at MEO orbit? at GEO orbit? (See "Problem Background" section for the answers.)

Understanding the fundamental concepts will facilitate your future learning processes. Good luck!

Linda K. Lanham
Structural Engineering

In high school I studied algebra, geometry, and trigonometry, but I was surprised to find myself needing these skills frequently when using drafting tools on computers. If you don't know geometry and trigonometry, you cannot really draw with the computer drafting tools. In fact, I often use a piece of scratch paper to figure out how to draw what I need on the computer, which was also not what I expected.

At my 9th grade science fair, I won a gift certificate from a drafting supply company. This started my interest in drafting, and resulted in several more scholarships. One was from a science fair, and another was the four-year General Electric National Scholarship from the Society of Women Engineers. I chose to attend an engineering college, the Colorado School of Mines.

After two years, I was getting a bit tired of school, so I accepted a cooperative education work assignment with the U.S. Forest Service. It was so much fun that I continued for three more years. I worked with older people, completed assignments, grew up, and gained experience towards a career, without knowing how it would work out. During my junior year, I also took one semester off to work fulltime with the U.S. Forest Service.

When I graduated with a BS in Mining Engineering, I went to work for the U.S. Forest Service, and have stayed with them ever since. I was an inspector of road and bridge projects and a road designer of low-volume roads in Colorado and Idaho. Now I am a bridge designer for the U.S. Forest Service Northern Region in Missoula, Montana.

Low-volume roads are unpaved timber and fire roads. The bridges I design are made of treated timber and concrete. I make a pre-design site visit, go back to my office to design the bridge, inspect the work in progress, and make a final inspection of the completed bridge.

One very easy math problem that many people find difficult is converting units—pounds per square inch, pounds over an area, volume, etc. In my work, I have to solve problems that involve conversions all the time. For example, a gabion is a rectangular wire cage filled with rock used in stream bank protection and headwalls. I measure its dimensions in feet and calculate its volume in cubic feet, but the material is ordered in cubic yards, not cubic feet. A simple error in conversion can translate into an expensive mistake.

Support for a Bridge

Timber is used to make bridges on unpaved fire roads through a forest. To build a bridge, calculate how large the wooden posts must be to hold the weight of the bridge and trucks that use the bridge.

Use the formula $F_c = P/A$ where:

$$F_c = \text{compressive stress or force; units lbs/in}^2$$

$$P = \text{force or load; units lbs}$$

$$A = \text{area; units in}^2$$

Problem 73. Determine the minimum dimensions of wooden post and steel-bearing plate when loaded with 100,000 lbs. (Ignore weight of wood post and bearing plate.)

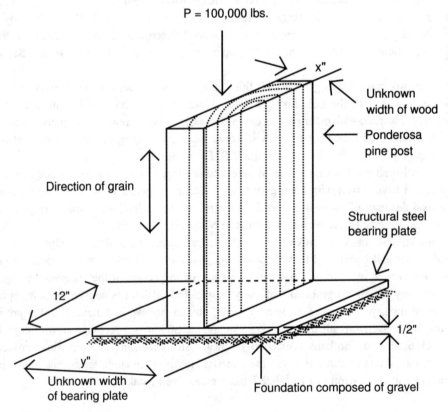

FIGURE 29
Wooden post and steel bearing

Step 1. Check compression parallel to grain—maximum crushing strength for the ponderosa pine post resting on the bearing plate. The mechanical properties of ponderosa pine for compression parallel to grain is 2450 lbs/in² (maximum). Given the force of compression, the total load, and the width of the board in one direction (12″), determine x, the unknown width of the wood.

Step 2. Check the compressive strength of gravel foundation material and the bearing plate. The ultimate bearing pressure for gravel is eight tons/ft². A good design for foundation bearing pressure is 2.0. Allowable bearing pressure, or compressive strength, for the gravel is four tons/ft².

First, convert the allowable bearing pressure for the gravel from tons/ft² to lbs/in². Next, given the bearing pressure for the gravel, the total load, and the width of the bearing plate in one direction, determine y, the width of the plate in the other direction. If the bearing plate was square, instead of 12 inches on one side and y inches on the other, what would its width need to be?

Step 3. Check the bearing plate's resistance to punching shearing stress due to the loaded wooden post. The ultimate shearing stress of structural steel is 25,000 lbs/in². From Step 1, the stress between the wooden post and bearing plate is Fc = 2450 lbs/in². Looking at one square inch of the steel material:

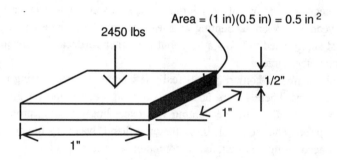

FIGURE 30
Steel material

The diagram shows that the area on the edge of the square inch is 0.5 inches, since the plate is 1/2 (0.5) inches thick. The formula for shearing stress, F_{shear}, is the same as F_c. Given the total force on the square inch, 2450 lbs, and the stress area, 0.5 inches, determine the shearing stress for the plate. Since the ultimate shearing stress of structural steel is 25,000 lbs/in, is the 1/2-inch plate adequate? Is a thicker plate necessary, or would a thinner plate do just as well?

Marla Parker

Computer Science

Every year I was in school, I studied math. Fortunately, my school, Baton Rouge Magnet High School, offered advanced-placement calculus to seniors. So, instead of learning calculus in a class of a hundred or more college freshmen, I was in a class of fewer than 30 high school seniors, with an enthusiastic teacher.

After high school, I went to Rice University in Houston, Texas. Since I grew up in Houston and lived in Baton Rouge for only my last two years of high school, going to Rice was an odd mixture of leaving home for college and going back home to Houston. At Rice, I was not expected to declare a major until my junior year, so I studied science and engineering. In my sophomore year, I took an accounting course—and loved it. Since I thought that modern accountants must surely need to know a lot about computers, I decided to major in accounting and minor in computer science.

In the middle of my sophomore year, I really left home, by transferring to the University of California, Berkeley. There I took another accounting course and an introductory Pascal programming class. This time, the accounting was exceedingly boring, probably because the teacher was uninspired. The Pascal class, on the other hand, was very fun and easy. I loved it, so I forgot about accounting and majored in computer science instead.

I received a BA in computer science in December, 1983. For over two years, I worked for a small computer-aided engineering company, and then went to work for Sun Microsystems, a company that manufactures workstations and "enterprise" computers, also known as servers.

At Sun, I've had three different jobs, but in all of them I have done basicly the same thing: write computer programs. For one-time small jobs, sometimes I write programs called "scripts." The first set of problems is based on one of my small scripts.

In 1991, I earned a private pilot license, with an airplane, single-engine, land rating. In 1992, I added an instrument rating, which means I can fly in the clouds instead of just under or around them, as before. Learning to fly under Visual Flight Rules is fun and pretty easy. Under Instrument Flight Rules, learning to fly is even more fun, but much more difficult. The most fun part of IFR flying is popping in and out of small, fluffy white clouds on an otherwise sunny day, or skimming in and out of the tops of clouds in the bright sunshine, while the earth below is all gray under the cloud layer. The second set of problems is based on a procedure that may be required when flying under Instrument Flight Rules.

These two sets of problems have one thing in common: logic. To solve them both, ordered, logical thinking is required. Since all of computer science is problem-solving, the reasoning skills that develop through studying and understanding technical subjects, like math, are the key to success. The same kind of reasoning skills are necessary for aviation, too.

Formatting Customer Data

My job requires collecting customer information data from several sites around the world, and formating the data so it can be read into a new database. The sites have been using an old program that saved the customer information in simple files, one per customer, rather than in a real database. We are switching over to a better program that will store the data in a database, where it will be much easier to access. To save the old data, it must be loaded into the new database.

The program that will load the old data into the new database expects the data to be in a certain format, different from the format of the old data files. So I need to write a script to convert the old data into the correct format.

The new format requires one field per line, seven lines per record. The fields are always in the same order: F-number, Name, Street, City-State, Zip, Phone, Fax. For example:

        F123-874
        Fred Wannabechi
        3807 Dryden Way.
        Frisky Kittens, UT
        80315
        801-555-3434
        801-555-3435

That would be the output produced from an old input file that looks like this:

        F-number: F123-874
        Customer: Fred Wannabechi
        Address: 3807 Dryden Way
        City-State: Frisky Kittens, UT
        Zip: 80315
        Phone: 801-555-3434
        Fax: 801-555-3435
        Comments: Random comments we will leave behind because
                they are not worth adding to the new database.

The input files also contain other data that we don't care about, but all the data we need is not always there. Sometimes the phone number is missing, or the address, or even the customer name. In these cases, the script must generate a blank line for the missing field. Each customer record must contain exactly seven lines, even if most of them are blank. The only field that is sure to exist is the F-number, because there is always exactly one F-number per input file. All the input files are joined together into one very long stream of input data for the script to process.

So, here is my first cut at the script. I wrote the script in a language called "awk" (perhaps because it is awkward to use), so here I'll use pseudo-code instead. Awk is an interpreter that processes input according to a script (the one I'm writing) one line at a time. That is, it executes the entire script for each line of input. The special symbols BEGIN and END mark code that will be executed only once, instead of once for each line of input. The BEGIN code is run before the first line of input, and the END code after the last line of input. If my input file has 100 lines in it, awk will execute my BEGIN code once, then the body of my script for each line of input, and finally my END code once.

The variables $0, $1, $2, et cetera are used to identify parts, or fields, of the current input line. A line is divided into fields by the field separator I am using, a colon ":". $0 means the entire line, $1 means the first field, $2 the second, and so on. For example, when the third input line in the sample input above is the current line, the variables would be:

$0 = Customer: Fred Wannabechi
$1 = Customer
$2 = Fred Wannabechi

Comments in the script, which will be ignored by awk, begin with a # sign.

```
  BEGIN code
# this code will happen only once, before reading any of the input
  set first = 1
  set FS = ":"
# FS is a special variable, the field separator. It defines the
# character used to define the $1, $2, etc. parts, or fields, within
# one input line
#
# By default FS is a blank, but I want it to be ":" so the
# labels in my input file will always be $1 and the data I
# want will be $2, the rest of the line after the ":"

  EACH LINE code
# this code will happen once for every line of input that is processed
# for every new line. The code will start running again right here and
# go to the END
  if ($1 = "F-number" ) then
      if (first = 1) then
          # initialize the fields to empty strings so that if an
          # input field is missing, a blank line will get printed in
          # the output in place of the missing field
          set first = 0
          set customer = " "
          set address = " "
          set citystate = " "
          set zip = " "
          set phone = " "
          set fax = " "
      else
          # Output the completed record for the previous fnum
          # This is the only place in the script that produces output
          print fnum
          print customer
          print address
          print citystate
          print zip
          print phone
```

```
        print fax
      #note the code here is part of the if ($1 = "F-number") then ...
      set fnum = $2
# note the code here is more EACH LINE code
    if ($1 = "Customer") then
        set customer = $2
    if ($1 = "Address" ) then
        set address = $2
    if ($1 = "City-State") then
        set citystate = $2
    if ($1 = "Zip") then
        set zip = $2
    if ($1 = "Phone") then
        set phone = $2
    if ($1 = "Fax") then
        set fax = $2
# note if $1 = anything else, the input line will simply be ignored
# this is the end of the EACH LINE code, go back to the beginning of
# the EACH LINE code to process the next line of input
END code
# this code happens once, after the last line of input has gone
# through the EACH LINE code but
# there is nothing to do here
```

To understand how the code works, you have to act as the computer. Do the BEGIN code, then read the input data and apply the EACH LINE code to it. Then do the END code (except in this program the END code section is empty aside from comments, so there is nothing to do at the end). While you are acting as the computer, you'll need a piece of scratch paper for notes to remember, and another piece of paper to write the output the program will produce.

Doing a Walkthrough

Pretending to be the computer and executing a program is called doing a walkthrough. The purpose of the walkthrough is to make sure your program has no bugs and will produce the correct output. When writing computer programs, this technique is used over and over again. Sometimes you can do the computer part in your head, but sometimes you need to make notes so you can remember what the computer would have stored in its memory. It is always a good idea to write down the output that the computer would generate, since that is the whole point of this sort of program.

The rest of this explanation takes you through a detailed walkthrough for the first few lines of input.(If you have done any programming or think you understand it well enough, you can skip this explanation and try the problems first.)

To walk through this code, you need to know how to do if-then and if-then-else statements. In the pseudocode above, I've used tabs to indicate the different parts of the if-then and if-then-else statements. If the "if" part of a statement is true, then you do the "then" part. If it is not true, you do the "else" part, or nothing if there is no "else" part. You never do both the "then" and the "else" parts of an if-then-else statement. After doing the if-then or the if-then-else parts, you do whatever comes next.

Do the BEGIN code by noting on your scratch paper that `first = 1` and `FS = ":"`. Then using the example for customer Wannabechi, the first line is:

```
F-number: F123-874
```

Remember that when you are in the EACH LINE code, the variables $0, $1, and $2 are set to mean the whole line, the first field, and the second field. So for the first line of input, the variables are set by the computer like this:

```
$0 = F-number: F123-874
$1 = F-number
$2 = F123-874
```

The first line of code in the EACH LINE code says:

```
if ($1 = "F-number" ) then
```

And in fact, for this first input line it is true that $1 = "F-number", so do what it says to do in the "then" part of the code. The first line of code inside the "then" part is another "if" statement:

```
if (first = 1) then
```

This "if" statment is also true, as you can see by checking your scratch paper that says `first = 1`. So do the "then" part also, by noting these variables and their values on your scratch paper:

```
set first = 0
set customer = " "
set address = " "
set citystate = " "
set zip = " "
set phone = " "
set fax = " "
```

Be sure to cross out the `first = 1` line on your scratch paper and replace it with `first = 0`.

Next, skip the "else" code, because you took the "then" code. But there is one more piece of code for the first if-then, after the end of the second if-then-else code:

```
set fnum = $2
```

So make a note on your scratch paper that `fnum = F123-874`, since the second field, $2, is F123-874 in the current input line.

The next line of code is:

```
if ($1 = "Customer") then
```

Since we know $1 = "F-number", not "Customer", this "if" statement is false, so you do not do the "then". Go to the next line. As it happens, all the rest of the "if" statements are also false, so don't do any of the remaining "thens". You are done with the first line of input.

Read the second line of input:

```
Customer: Fred Wannabechi
```

And go back to the first line of the EACH LINE code:

```
   if ($1 = "F-number" ) then
```

But for this line, this "if" statement is not true. $1 = "Customer", not "F-number", so skip this "then" statement. The next line of code is:

```
   if ($1 = "Customer") then
```

This is true, so do the "then" line:

```
   set customer = $2
```

You will write customer = Fred Wannabechi on your scratch paper.

The next "if" statement and all the rest are false for this line of input, so go to the next line of input.

Eventually, you will finish with this input, and start the next input. Then the "if" ($1 = "F-number") statement will be true, but first will be 0, not 1. So you will finally do the "else" statement, and print out all the values of the variables fnum, customer, address, et cetera, on the piece of paper you are using to show the output of the script.

Problem 74.   Make up some input data and run it through the script. See if it will produce the correct output, given the input you invented. Do one example that has all the required fields, then do an example with some fields completely missing.

Problem 75.   There are at least two bugs in my original version of the script. Bugs are mistakes that cause wrong or incomplete data to be printed as output. See if you can find the bugs.

Problem 76.   This problem and the next reveal what the two bugs are, so you can try to fix them. If you want to find the bugs on your own, stop reading now!

The first bug has to do with the requirement that a blank line be printed out in place of any field that is missing in the input. The output variables (customer, address, etc) are initialized to " " (a blank) when first = 1. If there are missing input fields in the first set of data, they will be correctly represented as blank lines in the ouput. However, what happens if the first set of data is complete? No fields are missing, so they are all printed out. But the second set of data has a missing address, so the address from the first set of data will be printed in its place. This is definitely a bug!

Rewrite the script to fix the bug.

Problem 77.   The second bug has to do with the last record. As originally written, the data for the last customer in the input file will never be printed out. That's because the print statements for one customer happen only after the script finds the F-number for the next customer. When there is no next customer, (because the script just processed the last one in the input), the print statements will never be executed.

Rewrite the script to fix this bug.

Entering a Hold

When you fly, the direction the plane is pointing is called your "heading". All aircraft have an instrument called the heading indicator. It looks something like Figure 31. The face of the heading indicator shows a "compass rose", including all the points on the compass. North

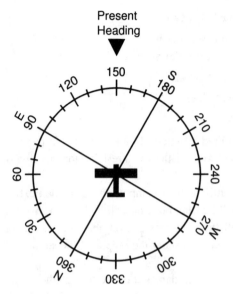

FIGURE 31
Heading indicator, while flying 150 = somewhat southeast

= 360°, South = 180°, East = 90°, West = 270°, Southwest = 225°, et cetera. Headings are usually expressed in degrees rather than directions, for example, "fly heading 180", rather than the less specific "fly south." As you fly, the heading indicator will turn so your current heading is always shown at the top.

When flying under Instrument Flight Rules (IFR) in the clouds, you cannot see anything out the windows except gray. You are in constant radio contact with an aircraft controller on the ground. The controller gives instructions like, "turn to heading 050, climb to 7000." You read the instructions back to make sure you heard correctly, and then comply immediately.

Sometimes, the controller tells you to wait awhile, before giving clearance to continue on your route or land at an airport. Since airplanes cannot stop in the air, like cars at red lights, the controller may instruct you to fly in a "holding pattern." A holding pattern is shaped like a racetrack in the sky. The starting gate of this imaginary, invisible racetrack is a "fix," usually a navigational radio station on the ground below. (Maybe it is called a fix because it is fixed solidly to the ground.)

The holding pattern is defined by a "radial," a heading that radiates out from the fix. Unless otherwise specified, all turns in the hold are to the right. In Figure 32, as you fly towards the

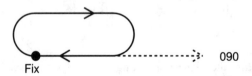

FIGURE 32
A holding pattern on the 090° radial from the fix

Fix, you are on the "inbound leg." After turning right, when you fly in the opposite direction on the otherside of the imaginary racetrack, you are on the "outbound leg."

Here are some examples of holding instructions and patterns. The holding patterns are drawn as if 360° (North) is at the top of the page.

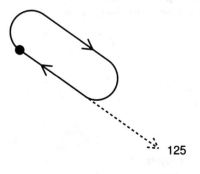

FIGURE 33
Hold on the 125 radial

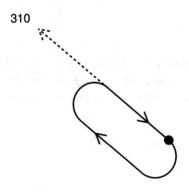

FIGURE 34
Hold on the 310 radial

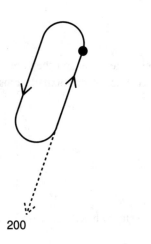

FIGURE 35
Hold on the 200 radial, left turns

When instructions to hold are given, the pilot must first turn towards the fix (if it isn't already directly ahead) and then figure out what type of entry to make into the hold: direct, parallel, or teardrop. To enter the hold properly, the type of entry must be determined first. Sometimes this requires quick thinking.

The direct entry is the easiest. Coming anywhere from the right of the heavy solid line, simply fly directly to the fix, and enter the hold by turning to the right. When you come from the left of the heavy solid line, you cannot do this easy direct entry, because the turn would be too sharp.

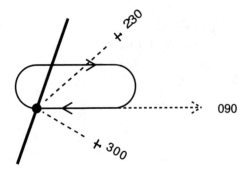

090

FIGURE 36
Direct entry to hold, right turns

The next-easiest entry is the teardrop, made from anywhere within the heavy V. After crossing the fix, turn to a heading of 30 towards the inside of the holding pattern, fly one minute, then turn right to fly inbound back to the fix, and continue in the holding pattern.

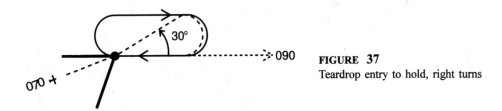

FIGURE 37
Teardrop entry to hold, right turns

The most difficult entry is the parallel, because it requires more turns and they are backwards. Notice that you need to turn left twice along the dashed line, before actually crossing

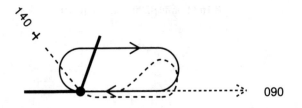

FIGURE 38
Parallel entry to hold, right turns

the fix in the correct direction to enter the hold. From then on, all turns are to the right, not the left.

When flying towards the fix, you will rarely have time to draw nice little pictures on the map and figure out how you are approaching the hold. Instead, there are rules that tell what kind of entry you are supposed to make, using the heading indicator. The rules are based on your current heading as you fly towards the fix, and the radial from the fix that defines the inbound leg of the hold.

As you fly towards the fix where you will enter a standard hold (one with all turns to the right), look at the compass rose to see where the heading for the radial that defines the hold falls. If it falls in area A, your entry should be teardrop. If it falls in area B, your entry should be direct. If it falls in area C, your entry should be parallel. If the radial falls on one of the dividing lines between two areas, use the easier of the two holding entries.

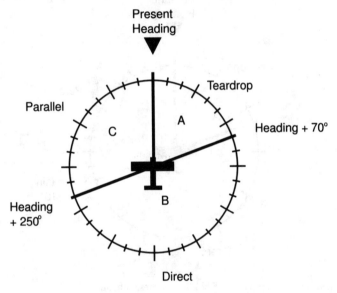

FIGURE 39
Compass rose divided up for three types of holding entries, right turns

Problem 78. Figures 32, 36, 37, and 38 all depict the same holding pattern, based on the 090 radial from the fix, with standard right turns. The headings of the planes shown in Figure 36 are 230 and 300. Using Figure 39 as a pattern, draw a heading indicator for the plane that is flying heading 230. See if the 090 mark falls in area A, B, or C. If you draw it correctly, 090 will fall in area B, indicating the direct entry shown in Figure 36. Do the same exercise for the other plane in Figure 36; its heading is 300. Then do the same exercise for the plane in Figure 37; its heading is 070. Finally, do the same exercise for the plane in Figure 38; its heading is 140.

Problem 79.   Using the holding pattern shown in Figure 33, what type of entry should you make if you are flying towards the fix, and your present heading is 220? 030? 100? Note that the radial that defines the holding pattern in Figure 33 is the 125 radial.

Problem 80.   Using the holding pattern shown in Figure 34, what type of entry should you make if you are flying towards the fix, and your present heading is 180? 300? 360? Note that the radial that defines the holding pattern in Figure 34 is 310 .

Problem 81.   Figure 35 shows a non-standard holding pattern. All the turns are to the left, not the right. The rules shown in Figure 39 on the compass rose will not work for a hold where all turns are to the left. Instead, you must use this pattern:

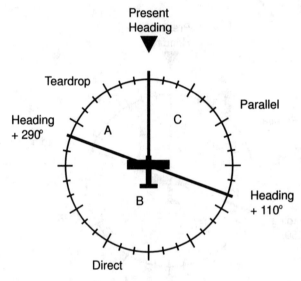

FIGURE 40
Compass rose divided up for three types of holding entries, left turns

One way to remember which pattern to use: For the standard right turns, the teardrop area A is on the right side. For left turns, the teardrop area A is on the left.

Using the holding pattern shown in Figure 35, and the compass rose shown in Figure 40, what type of entry should you make if you are flying towards the fix and your present heading is 250? 120? 020? Note that the radial that defines the hold shown in Figure 35 is 200.

Eileen L. Poiani

Mathematics

My Nutley High School yearbook caption says that I "would like to teach college math," probably because I had taken five year-long mathematics courses, plus the usual college prep curriculum. But why did I choose—and remain with—math? No doubt for a variety of reasons: the influence of my father's engineering background; my mother's competence in basic consumer mathematics; the American commitment to mathematics and science in the aftermath of the Soviet Union launching the spacecraft, Sputnik; and the fact that I liked mathematics—and find it a continuing challenge.

I fulfilled my career goal by entering Douglass College, the women's college of Rutgers University in New Jersey, majoring in mathematics with a minor in French. Then, I went directly into Rutgers' graduate school under a fellowship, and completed my PhD in mathematics six years later. As a full-time faculty member for three of those six years, I had to make every minute count to meet all the responsibilities of both a teacher and a student. I put many miles (and even more kilometers!) on my car, consulting with my dissertation advisor who was taking a year's sabbatical in a distant city.

Fortunately, I have always been a well-organized person so I was able to juggle both roles, but it did mean sacrificing a good deal of personal time.

My field of research is in real analysis and special function theory, part of pure mathematics. Currently, my research interests focus on improving the teaching and learning of mathematics; engaging those who are under-represented in mathematics; curing math avoidance and anxiety; and strategic planning in higher education.

I was the first woman mathematics teacher at Saint Peter's College, an all-male Jesuit institution that became coed soon after I joined the faculty. Nearly three decades later, I am one of two women in a department of eleven full-time members, after working my way up the academic ranks to the position of Professor of Mathematics.

In addition, I'm Assistant to the President for Planning, applying my mathematics background to a variety of administrative responsibilities, from strategic planning to accreditation and institutional research. In my experience, good communication and interpersonal skills are as essential as mathematics for both administration and teaching.

My involvement in professional mathematics activities began when I was elected Secretary-Treasurer of the New Jersey Section of the Mathematical Association of America (MAA). That led to my election as Governor of the Section, and to many other exciting roles, including President of Pi Mu Epsilon (the honorary national mathematics society), and Chair of the United States Commission on Mathematical Instruction, created by the National Research Council of the National Academy of Sciences.

My first real experience with a computer was during my summer job as a programmer at Bell Laboratories in 1964, where the computer nearly filled a room. I didn't use a calculator until several years later. But in seventh grade, I remember my science teacher telling us that, when we grew up, we would be as comfortable using a computer as a telephone. That prediction came true. Today, the calculator and computer are incredible resources for the mathematician. Imagine how thrilled I felt many years later, when I met Admiral Grace Murray Hopper, mathematician and computer pioneer, and learned first-hand about her role in the development of the computer.

Many young people, especially young women, drop out of high school math after only one or two years. I began encouraging them to persist in mathematics and to keep all their career doors open—whether that meant going to college or entering the job market. This led to my serving as the founding national director of *WAM: Women and Mathematics,* a lectureship program begun in 1975 and still growing. WAM is sponsored by the MAA under grants from IBM and other corporations.

I enjoy speaking to middle- and high-school students about solving mathematical problems. Some of my favorite topics are drawn from topology because they have interesting applications. I have taught a course in undergraduate topology and hope this example will influence you to study topology further.

What is Topology?

Topology is a branch of mathematics whose foundation is drawn primarily from geometry, set theory, and analysis. The complexity of modern day topology defies any clear, concise definition. However, the word "topology" is derived from the Greek word, topos, meaning "place", so in a sense topology is the study of places—and strange places indeed.

Topologists, like most mathematicians, are specialists. Their areas of specialization include point set theory, plus algebraic and differential topology. The elementary topics of topology and the notions of network theory fall within the domain of the so-called "point set" branch of topology. Early developments in point set topology can be traced to the works of Leonhard Euler (1707–83), August Möbius (1790–1868), Felix Klein (1849–1925), and Henri Poincaré (1854–1912). The establishment of a formal theory of sets by Georg Cantor (1845–1918) gave strong impetus to the growth of topology. Nevertheless, it is only within the last 75 years or so that topological theory has been formalized and its importance recognized. Areas of application for this 20th-century discipline range from celestial mechanics, electronics, and chemistry, to transportation, economics, and city planning.

Early in its development, topology was called "rubber-sheet geometry." Intuitively, this title refers to visualizing a curve drawn on a flexible piece of material that can be stretched, shrunk, or distorted, but never torn. The topologist studies various properties of the new curve in relation to the original. The Euclidean geometry notions of area, circumference, shape, congruence, similarity, and so on, are no longer applicable to this rubber-sheet geometry. Euclidean properties are based on rigid motions (i.e., rotations, reflections, inversions, and translations), while topological properties are based on *elastic* motions. Two figures are said to be "topologically alike" or homeomorphic, if one figure can be transformed into the other by a series of elastic motions. For instance, to a topologist, the following sets of figures are topologically equivalent:

FIGURE 41
Circle equivalents (A), polyhedron equivalents (B)

Problem 82.   Königsberg Bridge Problem

This is the classical problem of topological network theory. It was posed by Leonhard Euler (1707–1783) in the *Proceedings of the St. Petersburg Academy of Sciences* in 1736.

In the 18th century, seven bridges joined the island and main part of the city of Königsberg, Russia. (The name had originally been Kaliningrad.) The city is situated where the Old and New Pregel Rivers meet to form the main Pregel River. The problem states:

> Would it be possible to take a walking tour of Königsberg, pass over each of the seven bridges exactly once, and return to the starting point?

The schematic drawing of the bridges is:

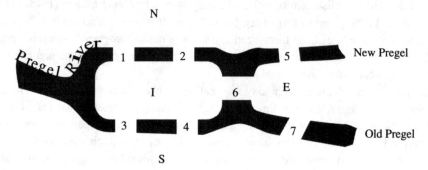

FIGURE 42
Königsberg bridges

As you can see, Königsberg surrounds the waterways and includes the island.

The layout of this city can be represented "topologically" by shrinking each section of the city to a point, and connecting the points with arcs that represent the bridges, as shown in Figure 43. Try to walk around the city, beginning and ending at the same point.

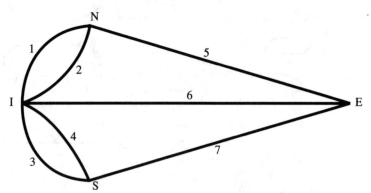

N = section of city north of the river E = section of city between New and Old Pregel River
S = section of city south of the river I = island section of city

FIGURE 43
Graph representing city of Königsberg

Network of Roadways

Problem 83. Consider the network of roadways in Figure 44. What is the minimum number of paths needed to traverse this network? (That is, to cover each arc once and only once.) Is it possible to traverse the network with an Euler circuit? (An Euler circuit is a single path that traverses the entire network, beginning and ending at the same point.) Indicate the paths (or path) on the diagram.

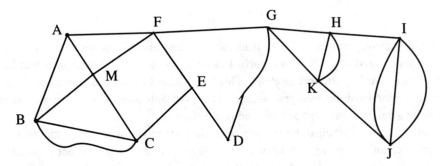

FIGURE **44**
Network of roadways

House Floor Plan

Problem 84. In the given floor plan, could you take a walking tour of the house (including the outdoors), pass through each door once and only once, and return to the room in which you started? Why or why not?

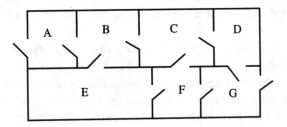

FIGURE **45**
House floor plan

Nancy Powers Siler

Dietetics—Foodservice Management and Nutrition

In high school, I was in advanced math, science, and English classes until my senior year, when I became tired of studying so much. I switched to an easier course load that included economics, regular English, and electives. This way, I could continue my social activities, and became a Senior Favorite upon graduation. I was even Salutatorian, with the second-highest grade point average in my high school graduating class.

However, when I reached the University of Houston in Houston, Texas, and took general and organic chemistry, along with a few other difficult courses, I regretted not continuing with trigonometry and chemistry in high school. Being Senior Favorite did not help me in my studies, but I managed to graduate with a Bachelor of Science Degree in 1971 after only three and a half years in college. During those years, I was also working a minimum of twenty hours per week in a hospital business office, using math to calculate patient bills, insurance reimbursements, laboratory and drug charges, and general accounting assignments.

At St. Paul Ramsey Medical Center in St. Paul, Minnesota, I completed a Dietetic Internship. There I used math on a daily basis to assess patient's nutritional status and calculate diet prescriptions. Next, I successfully completed a National Registration Examination given by the American Dietetic Association Commission on Dietetic Registration, and became a registered dietitian.

In my first two professional jobs, I was a clinical dietitian, which involved calculating nutritional needs and counseling patients with their diets. I worked with heart, renal, gastrointestinal, pulmonary, diabetic, general medical, and surgical patients. For one fun project, a cookbook, I developed recipes that were modified to reduce fat and salt content, yet still tasted good.

In order to pay for graduate school, I started my own consulting business, contracting with hospitals, nursing homes, child care facilities, drug treatment centers, and anyone else who would pay me. Taking the Graduate Record Examination made me again regret that I had skipped trigonometry in high school. However, I passed it on the first try, and went on to complete a Master of Science Degree from Southwest Texas State University in San Marcos, Texas. I graduated in eighteen months, with a 4.0 GPA, while working, being married, maintaining an active social life, and running a household.

I expanded my business by working with architects, designing and remodeling new and existing hospitals and nursing homes. Math is an integral part of layout and design, in addition to costing projects. One small calculation error and the ovens won't fit in the new bakery, or the walkway between work centers won't meet state or national safety code standards. Then, the foodservice operation can't open! This work experience helped me provide insight to the real world to my students at The University of Texas at Austin, Southwest Texas State University (former President Lyndon Baines Johnson went there, but not when I was teaching), and Chicago State University, when I became a teacher later on.

After fourteen plus years running my own business in Texas, I followed the love of my life to the Chicago area, and currently hold the position of Coordinated Program Coordinator for the Department of Nutrition and Medical Dietetics in the College of Associated Health Professions at the University of Illinois at Chicago. I teach students how to calculate nourishing formula feedings that can be administered to patients by nasogastric (nose to stomach), jejunostomy (directly into this section of the intestine), or central vein (directly into the bloodstream) routes. Also, I show them how to compute the vitamin and mineral content of a daily diet, and how to compare it to what a body needs to function properly. I also use math to determine student grades each semester, keep up with my children's swim times, help them with their homework and school projects, plus count the hamsters which keep multiplying at our home.

Yield Problems

In foodservice management, math is an integral part of the daily work schedule. The problems given below are practical examples of what a foodservice director and staff must do to purchase food and prepare meals for their clients, whether they are in a school, fast food establishment, healthcare facility, or an upscale restaurant.

3 teaspoons = 1 tablespoon
2 tablespoons = 1 fluid ounce
8 fluid ounces = 1 cup
2 cups = 1 pint
2 pints = 1 quart
4 quarts = 1 gallon
16 ounces = 1 pound

The index number(#) on a dipper or ice cream scoop indicates the number of servings per quart. For example, a level #10 scoop provides ten servings per quart. Since there are 32 ounces per quart, each serving would be about 3.2 ounces.

Problem 85. Convert the following measures:
A. 46 ounces = _____ pounds
B. 3/4 cup = _____ tablespoons
C. 1/2 gallon = _____ cups
D. 3 ounces = _____ teaspoons
E. 6 1/2 quarts = _____ ounces
F. 2 gallons = _____ pints
G. The recipe for pimento-cheese sandwich filling yields 2 1/2 gallons and is portioned for sandwiches with a #24 scoop. How many sandwiches can you make from this recipe?

    a. 320   c. 240
    b. 80   d. 380

H. What size scoop should be used to portion approximately one ounce of filling per sandwich?

    a. #16   b. #20
    c. #24   d. #30

I. A standardized recipe for chicken salad yields four gallons and is portioned for sandwiches with a #20 dipper. How many sandwiches can be made from this recipe?

    a. 320   b. 80
    c. 240   d. 380

J. If there are approximately 28 slices of bread per two-pound Pullman loaf of bread (sandwich-style or square bread), how many two-pound loaves of bread sliced 1/2 inch thick are required to prepare 100 ham sandwiches?

    a. 9   b. 7
    c. 6   d. 4

K. There are 32 tablespoons of butter in a pound. How many pounds of butter are needed to butter both slices of bread for one hundred sandwiches, using approximately 1 1/2 teaspoons per slice of bread?

      a. 1.5 pounds   b. 2.0 pounds

      c. 2.5 pounds   d. 3.0 pounds

    L. It takes one teaspoon of mayonnaise per slice of bread for turkey sandwiches. To prepare 500 whole sandwiches, how many gallons of mayonnaise are needed?

      a. 1/2 gallon   b. 3/4 gallon

      c. 1 gallon   d. 1 1/3 gallons

Assestment of Food Intake

The duties of a clinical dietitian include analyzing the food intake of the clients served. The following activities are an example of what might occur during a counseling session.

Problem 86.   A. Twenty-four Hour Recall. Write down the foods eaten or beverages consumed in the last 24-hour period, including the amount (1/2 cup, 4 ounces) and type of preparation (fried, baked, battered). Include everything that has entered your mouth including water, gum, etcetera.

    B. Guide to Good Eating. Compare your 24-hour recall with the National Dairy Council's Guide to Good Eating. Categorize the foods you eat, and total the number of servings in each group. It is best to eat more servings from these food groups rather than foods from the "other group," since the listed food groups frequently provide more nutritional benefits.

	My Intake	Recommended for Teenagers
Milk Group	____ Servings	4 Servings
Serving size examples:		
1 cup milk or yogurt		
1 1/2 oz cheese		
1 cup pudding		
1 3/4 cup ice cream		
2 cups cottage cheese		
Meat Group	____ Servings	2 Servings
Serving size examples:		
2 ounces cooked lean meat,		
fish, poultry, or		
2 eggs		
2 ounces cheese		
1/2 cup cottage cheese		
1 cup dried beans or peas		
4 Tbsp peanut butter		
Fruit-Vegetable Group	____ Servings	4 Servings
Serving size examples:		
1/2 cup cooked fruit or vegetable		
1/2 cup juice		

       1 cup raw fruit or vegetable
       Medium-sized apple
       Medium-sized banana
       Medium-sized orange

Grain-Starch Group   ____ Servings   4 Servings
Serving size examples:
       1 slice bread
       1 cup uncooked cereal
       1/2 cup cooked cereal, pasta,
          rice, grits, etc.

Other Foods   ____ Servings   None

By selecting lean meats that are baked or broiled instead of fried, along with vegetables and starches that are unbuttered or not creamed, you can eliminate fat and calories. By eating skimmed or non-fat milk, yogurt, and cheese, you can also lower your fat and calorie intake, adding many nutritional benefits to your overall health.

Estimation of Weight, Energy, Protein, and Fluids

Another task of a clinical dietitian in community clinics, corporate wellness centers, hospitals, nursing homes, drug treatment centers, or in private practice, would include determining the correct weight, energy, protein, and fluid needs of the clients.

Weight. If height/weight tables are not available, use the "rule of thumb" quick estimate:

Men—106 pounds for the first five feet of height, plus six pounds for each inch over five feet.

Women—100 pounds for the first five feet of height, plus five pounds for each inch over five feet.

Calorie or Energy Needs. Basal Energy Expenditure (BEE) is the energy required for basic life processes such as breathing, heart functioning, and maintenance of proper body temperature. It can be determined using the following formulas:

Men—BEE = 66 + (13.7 * Weight in kilograms) + (5 * Height in Centimeters)
      − (6.8 * Age in Years)

Female—BEE = 655 + (9.6 * Weight in kilograms) + (1.7 * Height in Centimeters)
      − (4.7 * Age in Years)

An adjustment must be made for the calories expended in daily activities such as walking, eating, etc. A person who repairs electrical equipment at the top of telephone poles performs more strenuous work than someone who answers the phone all day. The first person burns more calories, so their calorie requirement should reflect that. Multipy the BEE figure by the activity factor to obtain the total calories needed for the day.

    Adjustment For Activity:
        Sedentary Activity = BEE * 1.3
        Moderate Activity = BEE * 1.4
        Strenuous Activity = BEE * 1.7

Protein Needs. Estimate by using the quick method of 0.8 to 1 gram of protein times the weight in kilograms for a healthy individual, or 1.2 to 2 grams of protein for a malnourished person.

Fluid Requirements.
> 18 to 55 years old = 35 cc per kilogram of body weight
> 56+ years old = 30 cc per kilogram of body weight
> Additional information: 240 cc = 8 oz
> Determining Weight in Kilograms: Divide weight in pounds by 2.2
> Determining Height in Centimeters: Multiply height in inches times 2.54

Percent Ideal Body Weight or Percent Usual Body Weight. To determine percent of ideal body weight or percent of usual body weight, use the following formulas:

$$\text{Percent Ideal Body Weight} = \frac{\text{Current Weight}}{\text{Ideal Body Weight}}$$

$$\text{Percent Usual Body Weight} = \frac{\text{Current Weight}}{\text{Usual Weight}}$$

Problem 87. A. Jim is a 35-year-old accountant who weighs 185 lbs and is 6′1″ tall. He runs four miles per day four times a week, and lifts weights two days a week. Determine his ideal body weight (IBW), and his calorie, protein, and fluid requirements. How many eight ounce glasses of water should he drink per day?

B. Jane is a 30-year-old secretary who weighs 135 lbs and is 5′5″ tall. Estimate her IBW and her calorie, protein, and fluid requirements. How many eight ounce glasses of water should she drink per day?

C. Jane's sister Sara is a 22-year-old teacher who weighs 165 lbs and is 5′7″ tall. Estimate her IBW and her calorie, protein, and fluid requirements. How many eight ounce glasses of water should she drink per day?

Problem 88. A. Evaluate the following weight for height:
Female   IBW: ____
Height 5′5″   Current weight: 128 lbs   %IBW: ____
         Usual weight: 145 lbs   %UBW: ____

B. Evaluate the following weight for height:
Male   IBW: ____
Height 6′0″   Current weight: 150 lbs   %IBW: ____
         Usual weight: 170 lbs   %UBW: ____

Problem 89. Estimate the recommended weight, calorie, protein, and fluid levels for the following people.
> A. Male, 27 yrs, 5′11″, 200 lbs
> B. Female, 33 yrs, 5′3″, 130 lbs
> C. Male, 81 yrs, 5′10″, 180 lbs
> D. Female, 83 yrs, 5′2″, 110 lbs

Problem 90.   Mrs. Green is a 38-year-old computer scientist who is 5'7" tall and weighs 150 lbs. Her usual weight is 175 lbs. Estimate her IBW. Determine her %IBW and %UBW.

Diabetes Mellitus

(See the Estimation of Weight, Energy, Protein, and Fluids section for weight, calorie of energy needs, protein needs, determining weight in kilograms, determining height in centimeters.)

Percentage of Calories.   To determine the percentage of calories from the calorie-producing nutrients carbohydrate, fat, protein, or alcohol use the following formula: percent of calories times the total calories divided by 100, or calories times percent of calories.

Example. In a 1500-calorie diet, how many fat calories are there if the diet is 30% fat? Multiply 1500 calories times 30 (percent) and divide by 100 to get the answer of 450 calories.

Calculating Grams of Carbohydrate, Fat, and Protein from Calories.   The physiological fuel-value of carbohydrate and protein is four. The physiological fuel-value of fat is nine. Take the number of calories for the nutrient and divide by the physiological fuel-value.

Example. Carbohydrate calories are 400. To find how many grams of carbohydrate that would yield, divide 400 by four to get 100 grams of carbohydrate.

Fat calories are 450. To find out how many grams of fat that would allow, divide 450 by nine to get 50 grams of fat.

Diabetic "exchanges" are listed in the left-hand column of the following chart. Exchanges are foods grouped together on a list according to similarities in food values. A single exchange contains approximately equal amounts of calories, carbohydrates, proteins, fats, minerals, and vitamins. For example, one fruit exchange equals one small apple, 12 grapes, 1/2 cup orange juice, or one medium tangerine.

Exchange List Nutritional Information

	Carbohydrate	Protein	Fat
Lean meat	0 grams	7 grams	3 grams
Starch	15 grams	3 grams	0 grams
Vegetable	5 grams	2 grams	0 grams
Fruit	15 grams	0 grams	0 grams
Fat	0 grams	0 grams	5 grams
Skim milk	12 grams	8 grams	0 grams
Lowfat milk	12 grams	8 grams	5 grams

Problem 91.   Case Study: Mrs. Frank

Mrs. Frank is a 45-year-old female with Non-Insulin Dependent Diabetes Mellitus. She is 5'4" tall, weighs 150 pounds, and has a small frame. She is moderately active.

A. What is Mrs. Frank's ideal body weight?

B. What is her basal caloric requirement using the Harris Benedict BEE?

C. What are her total kilocalorie requirements?

D. What kilocalorie level would enable her to lose one pound of weight per week? (3500 kilocalories = approximately one pound of body weight. To lose one pound per week, subtract 500 calories per day from the total kilocalorie requirement.)

E. How many grams of carbohydrate, protein, and fat should be included in her meal plan, using the guidelines of 15% protein, 60% carbohydrate, and 25% fat?

F. If Mrs. Frank consumes three meals per day, how many kilocalories and grams of carbohydrate should be provided at each meal? (Carbohydrates should be equally divided between the three meals.)

Problem 92.   Using the nutrient values of exchange lists provided, calculate the amount of carbohydrate, protein, fat, and total kilocalories provided in the daily allowance.

The following chart represents the number of servings allowed for each exchange in a particular meal. For example, breakfast includes one lean meat exchange (one ounce lean ham); two starch exchanges (one slice toast and 3/4 cup dry cereal, or two slices toast); no vegetable exchange; one fruit exchange (1/3 cup cranapple juice or one orange); one fat exchange (one teaspoon oleo or one slice of bacon); and one lowfat milk exchange (eight ounces lowfat milk).

Since the previous chart tells you that one fruit exchange provides 15 grams of carbohydrate, you know that one fruit exchange from the breakfast meat for this pattern provides 15 grams carbohydrate, and three servings for the entire day adds up to 45 grams carbohydrate. Since there is no protein or fat for the fruit exchanges, the answers in that column for the fruit exchange would be zero.

	Breakfast	Lunch	Supper	Carbohydrate	Protein	Fat
Lean meat	1	3	2			
Starch	2	2	2			
Vegetable	—	1	2			
Fruit	1	1	1			
Fat	1	2	2			
Lowfat milk	1	1/2	1/2			
			Totals:			

Total Calories: _____

Problem 93.   A. Utilizing the guidelines of 55 to 60% carbohydrate calories, 12 to 20% protein calories, and 25 to 30% calories as fat, determine the diet prescription for 2000 calories.

_____ grams carbohydrate
_____ grams protein
_____ grams fat

B. Write a meal plan in exchanges for someone who requires a bedtime snack in addition to the three meals per day.

Remember that the breakfast-through-snack column represents the number of exchanges allowed, while the carbohydrate, protein, and fat columns represent the values provided from the meal and snack columns. This is the format registered dietitians use to calculate meal patterns and the corresponding nutritional value assigned to these patterns.

	Breakfast	Lunch	Supper	Snack	Carbohydrate	Protein	Fat
Lean meat							
Starch							
Vegetable							
Fruit							
Fat							
Skim milk							
			Totals:				

Total Calories: _____

Enteral Nutrition (Feedings Given by Tube)

Frequently, hospital or nursing home patients are not able to eat enough calories to meet their nutritional needs. When this occurs, they can be fed through a nasogastric tube (small flexible tubing placed down their nose into their stomach). There are also other locations for tube placement that require an enteral formula. Enteral formulas are liquid, about the consistency of an instant breakfast drink, and can provide all the nutrients a person needs without ever "eating" as we normally think of it.

The objective of this series of problems is to determine the amount of calories, protein, percent of U.S. Recommended Daily Dietary Allowance of vitamins and minerals for the average adult, and the total water provided by the formula. Since adequate nutritional status is related to positive nitrogen balance, calculating the nonprotein kilocalories-to-nitrogen ratio and the total kilocalories-to-nitrogen ratio is one way to indicate adequacy. In a health care setting, patients' nutritional requirements would be compared to the physician's orders, to determine whether the order provides the nutrients required by the individual.

Total Milliliters in 24 Hours: Multiply the number of milliliters(ml) every hour by 24 hours

Total Calories: Number of calories per ml times the number of ml per 24 hours.

Grams of Protein: Number of grams of protein in 1000 ml times the total number of ml per 24 hours divided by 1000.

USRDA for Vitamins and Minerals: 100 times the total number of ml per 24 hours divided by the number of calories required for 100%.

Water from Formula:

> 1 calorie-per-ml formulas provide about 85% water
>
> 1.5 calories-per-ml formulas provide about 77.6% water
>
> 2 calories-per-ml formulas provide about 57.7% water

Multiply the number of formula ml times the proper percentage to calculate the ml of water provided.

Calculating Grams of Nitrogen: Divide grams of protein by 6.25

Nonprotein Kilocalories-to-Nitrogen Ratio: Total calories provided minus the grams of protein times the physiological fuel-value of four gives the number of nonprotein calories. Nonprotein calories divided by the grams of nitrogen supplies the ratio.

Total Kilocalories-to-Nitrogen Ratio: Total calories divided by the grams of nitrogen.

Problem 94.   Calculate

Total ml in 24 hours

Total calories

Protein (in grams)

USRDA vitamins and minerals (percentage)

Water from formula (in cc)

Grams of nitrogen

Non-protein kilocalories-to-nitrogen ratio

The total calories-to-nitrogen ratio

for each of the diet orders given below.

A. Diet order: 100 ml Osmolite every hour.   Osmolite provides 1.06 calories per ml and 37.2 grams of protein in 1000 ml. 1887 ml provide 100% of the USRDA for vitamins and minerals.

B. Diet order: Jevity 1600 ML every 24 hours.   Jevity provides 1.06 calories per ml and 44 grams of protein in 1000 ml. 1321 ml provide 100% of the USRDA for vitamins and minerals.

C. Diet order: Enrich 240 ml four times a day.   Enrich provides 1.1 calories per ml and 39.7 grams of protein in 1000 ml. 1391 ml provide 100% of the USRDA for vitamins and minerals.

D. Diet order: Traumacal 50 ml every hour.   Traumacal provides 1.5 calories per ml and 83 grams of protein in 1000 ml. 2000 ml provide 100% of the USRDA for vitamins and minerals.

E. Diet order: Twocal HN 80 ML every hour, half-strength.   Twocal provides two calories per ml and 83.7 grams of protein in 1000 ml. 1900 ml provide 100% of the USRDA for vitamins and minerals. Half-strength means the formula is diluted with water, so it provides one-half of the nutrients.

F. Diet order: Vivonex Ten 100 ml, 3/4-strength every hour.   Vivonex provides one calorie per ml and 38.2 grams of protein in 1000 ml. 12000 ml provide 100% of the USRDA for vitamins and minerals. Three-fourths strength means the formula is diluted with water so it provides three-fourths of the nutrients.

Fahmida N. Chowdhury

Electrical Engineering

I was born in Bangladesh, and went to all-girls schools where it never occured to anyone that girls were *not* supposed to be good at mathematics and science. In ninth grade, we had to choose a "track"—either science or humanities. Once you chose a track, you couldn't switch back to the other track, because you would have missed too many courses and there was no way of making them up.

The science track included advanced mathematics, physics, inorganic and organic chemistry, biology, calculus, statics, and dynamics. Both tracks taught algebra and geometry, but the science students took more of both. In eleventh and twelfth grades, science students chose between advanced geography and biology. To go to medical school, you had to take biology. However, by that time I had decided to be an electrical engineer, so I chose geography.

Although my childhood dream was to become a writer or painter, or both, I thought that the safest bet was to become an engineer. At that time, the job market in Bangladesh guaranteed that an engineering degree would lead to a decent job (alas, it is not so any more.) Besides, I thought that as an engineer, I could still write or paint in my spare time, but just try to be a full-time writer and do engineering on the side!

Of course, at that time almost no women went into engineering, particularly electrical. I didn't know any women engineers, hadn't even read about any, but someone told me there were a few women in some engineering departments at the university. The fact that electrical engineering was supposed to be off-limits to women actually drew me to it even more strongly. Both my parents were extremely supportive and very serious about their children's education.

After my exit examination at the end of the twelfth grade, I applied for a government scholarship to study electrical engineering in the (then) USSR. I applied simply for fun, to see whether I would be selected, since there was room for only 10 people for the program. When I was chosen, nobody thought I would actually go to such a faraway country—but I did. My parents helped me prepare for the trip, but told me later that they were shocked to find me 'gone' one day.

I spent more than six years in Moscow, getting a Master of Science degree in electrical engineering, specializing in electrical machinery. The medium of instruction was Russian, and for a few years Russian became almost like my first language. There were very few women in my classes, although there were more in the department of electronics and the institutes of chemical engineering.

During some summer vacations, I made a long train journey across Europe to visit relatives in London, England. On the way, I went through Poland, (then) East and West Germany, and Holland or Belgium, depending on the train's route. I could break my journey at any point, spend the day in a city, and catch another train at night, getting off in the morning in another city. In this way, I visited Warsaw, Poland, and both East and West Berlin, in (then) East Germany. One time, I made a detour to spend a few days with a friend in Amsterdam, visiting art museums and walking around the city. Most people I met gave me a strange look if I told them I was an electrical engineering student.

In 1981, I went back to Bangladesh, and my first job was in the telecommunications industry there. Later that year, I joined the Bangladesh University of Engineering and Technology as a lecturer in the electrical engineering department, where I was one of the first two women faculty members. In 1983, I came to the USA to earn a doctorate at Louisiana State University (LSU) in Baton Rouge, Louisiana. There, I switched my field from power to control systems and received the PhD in December, 1988.

By this time I had been married for six years and also had a one-year-old daughter. Since I didn't want to move, I switched my field to do postdoctoral research in biophysical chemistry at LSU. This may appear unusual, but there can be a lot of overlap between seemingly disparate fields of science. Even though I officially belonged to a chemistry department, my research involved mathematical modeling, statistical analysis, and technical software development. After a couple of years, I started teaching at Southern University in Baton Rouge, while continuing my research in the field of control systems, focusing on neural networks.

Currently, I teach electrical engineering at the Michigan Technological University and do research in parameter estimation and fault detection in dynamic systems, using neural networks as a tool. In my teaching, I try to convey my enthusiasm for the subject to my students. Believe it or not, this is fun! I hope my students discover the fun and excitement of this profession. A solid mathematical foundation is very important for engineering education, and I like to tell my students that even when it seems useless, mathematics is aerobics and yoga for the brain.

Replacing old Telephone Sets

My first job was as a research engineer for a telephone and exchange manufacturing company in my home country, Bangladesh. My first assignment was to decide whether we could replace some of the very old and unreliable telephone sets used in rural railway stations with new sets, without having to replace the whole system. These sets were so old, nobody was manufacturing them any more. In fact, nobody remembered how to make them, because all the technical information and manuals had been lost. These telephone sets were quite different from the modern ones that we knew how to make. So we decided to modify the modern set to include the necessary properties to make it work with the old system. We also recommended that in future, when the railway company could afford it, the whole system should be replaced.

Along with two colleagues, I took apart the sample telephone set given to us by the railway company. We wanted to see if we could make an inductor coil to match the technical specifications of the one we found inside the set. To reproduce the coil, we would have to specify its resistance, inductance, number of turns, material and insulation. The following problem is about specifying the resistance.

First, we measured the resistance of the old coil. But there were three people in the team, and each of us got a slightly different value. Let these values be x_1, x_2 and x_3 Ohms (units of resistance). To ask the workshop to make a coil, we had to choose one value. We also had to calculate the sample variance of our measurements. If it is too large, the measurements must be repeated. The following problem will show you how to choose a value and calculate the sample variance.

(By the way, our design was successful. After the preliminary testing of the telephone sets, I left the company to join the Electrical Engineering Department at the University, but I learned later that the sets were manufactured and sold to the railway company.)

Note. The numerical values used in this problem are fictitious, simply because I don't remember the actual numbers. The problem is very general, and is applicable to any experiment involving measured quantities. Here are some definitions that will help you solve this problem:

The arithmetic mean (or average) of n quantities, $x_1, x_2, \ldots, x_n$, is given by:

$$\text{mean } x = \overline{x} = \frac{x_1 + x_2 \cdots + x_n}{n}$$

In the absence of other evidence, the mean is the best estimate of a quantity you are trying to measure. The more measurements you make, the closer the mean gets to the true value with a high probability.

Sample variance is calculated by:

$$\text{var } x = \frac{(x_1 - \overline{x})^2 + (x_2 - \overline{x})^2 + \cdots + (x_n - \overline{x})^2}{n - 1}$$

In general, the larger the sample variance, the less trustworthy the mean. Variance is a measure of the "spread" of the observed data about the mean value. Standard deviation is defined as the square root of the variance.

The following problem (part D) can be solved without calculus or by using calculus. Here is a hint for a non-calculus solution: First, show by completing the square that $f(x) = ax^2 + bx + c$, $a > 0$, has its minimum value at $x = -b/2a$. Then consider

$$\text{err } = f(u) = (x_1 - u)^2 + (x_2 - u)^2 + (x_3 - u)^2$$

For a calculus solution, try this: To obtain the minimum (or maximum) of a function, take its derivative, and find out what values of the argument make the derivative function go to zero. To check if any of these values is a minimum, take the second derivative of the function and evaluate it at that value. If the result is positive (negative), you have a minimum (maximum).

Problem 95.   A. Take three measured values and calculate the mean. The mean value should be specified as the resistance of the coil you are trying to make. Use $x_1 = 99.6$, $x_2 = 101.5$, $x_3 = 100.1$.
  B. Calculate the sample variance and standard deviation.
  C. If the sample standard deviation is less than one percent of the mean, write down the specification in the form:

$$\bar{x} \pm \text{ tolerance}$$

where tolerance $= 3 *$ standard deviation, and $\bar{x} =$ mean value.
  D. Suppose your friend says that the best estimate of a quantity (using a number of measurements) is given by the value that yields a minimum of the total squared errors. This certainly seems reasonable, so find the best estimate according to this criterion, and compare this result with your best estimate, the average, which you have already calculated.

Reliability Analysis

My research work involves designing power system fault-detection schemes. At present, I am working on the problem of correctly estimating the phase voltages and currents immediately after a power system short circuit. These values are needed to help calculate the fault location and decide which relays should trip. For a good system design, I also must calculate the probabilities of different types of short circuits and equipment failures. The following problem is a partial, simplified version of real-life reliability analysis.

Reliability of equipment is defined as the probability that it operates without failure. Failure probability is, obviously, the probability that it does not work. The sum of these two probabilities is one, because it will either work or not work.

Problem 96.   Suppose, in a given segment of power line, there are two transformers—Tl and T2. Tl is very old and has a reliability of only 89 percent. T2 is relatively new, with a reliability of 96 percent. Assume they are independent, so the failure of Tl does not influence the reliability of T2, and vice versa. Calculate the following probabilities:
  i.   Tl fails, but T2 works
  ii. T2 fails, but Tl works
  iii. Both Tl and T2 fail
  iv. At least one of them works

Problem 97.   An industry runs from 700 to 1700 hours (i.e., 7:00 a.m. to 5:00 p.m.). Its efficiency varies with the time of the day, x, in the following way:

$$\text{eff (in \%)} = f(x) = -x^2 + 18x + 12$$

Can you find the time of the day when efficiency is at its maximum? Also, find this maximum efficiency. Assume that x varies between 7.0 and 17.0, in keeping with the working hours.

Rosalie Dinkey
Chemical Engineering, retired

In high school I took all the math classes that were offered: both beginning and advanced algebra, geometry, and trigonometry. Although I planned to major in chemistry, I liked math so well that I switched to chemical engineering, because I thought I would use math more in engineering. In 1948, I received a BS in Chemical Engineering from the University of Minnesota.

For several years, I worked at General Electric, Sylvania, and Stanford Research Institute as an analytical and production chemist. I wanted to learn more math, so I went to San Jose State University and in 1968, earned an MS in Mathematics. Two of my sisters were teachers, and I thought I might like to teach math at a junior college. However, there were no open teaching positions at local junior colleges, so I taught high school algebra and geometry at a private girls' school instead.

I did not enjoy teaching so much as I had hoped, so after three years I returned to engineering, this time at Hewlett-Packard. In my last job before retirement, I was in charge of oxidation-diffusion furnaces in a semiconductor manufacturing facility. When building semiconductor "chips" from silicon, oxides are grown on them. They are then patterned with photolithography equipment to make the various devices: transistors, resistors, and so on. Dopants, such as phosphorous, are diffused into the circuits to make the desired electrical properties. At the end of the process, more oxides are grown to protect the circuits from the outside atmosphere.

All of my jobs have used mathematics, especially algebra and statistics. One of the most important things a technical education can do is teach students how to learn what they will need to know when they change from one field to another. The knowledge of solid-state physics and materials needed in my last job wasn't even dreamed about when I graduated from college in 1948.

Growing a Scratch-Protection Coating

Here is a real-life problem of the sort I dealt with when I was making semiconductors for Hewlett-Packard. A situation like this, requiring the engineer to plug values into a formula, is very common.

To manufacture integrated circuits (or "computer chips"), many copies of the same circuit are produced on a single piece of silicon called a wafer. Later the wafer is cut up, and the individual circuits are tested and packaged individually.

The silicon dioxide scratch-protection coating on a silicon wafer with integrated circuits contains phosphorus that combines with sodium and other contaminants present in the wafer, diffusing it from the atmosphere. The machine that produces the phosphorus-doped silicon dioxide is old, and the phosphorus content varies widely from run to run.

Problem 98. Last Wednesday these were the readings for weight-% phosphorus:

Run Number	Value (weight-% phosphorus)
1	2.63
2	2.72
3	2.47
4	2.55

A. What was the mean value, $\bar{x}$, (to 2 decimal places)?

B. What are the individual deviations from the mean, i.e., calculate $x - \bar{x}$, for each x.

C. What is the standard deviation of the sample? To calculate the standard deviation of this small number of readings, use the formula:

$$s = \sqrt{\frac{(x_1 - \bar{x})^2 + (x_2 - \bar{x})^2 + (x_3 - \bar{x})^2 + (x_4 - \bar{x})^2}{3}}$$

Problem 99. The standard deviation measures the spread of the data, calculated by what is called a root-mean-square formula. Suppose the values for the readings were as follows:

Run Number	Value (weight-% phosphorus)
1	1.09
2	3.06
3	4.10
4	2.12

$$\bar{x} = (1.09 + 3.06 + 4.10 + 2.12)/4$$

$$= 10.37/4$$

$$= 2.59 \text{ weight-\% phosphorus, as in the previous problem.}$$

What is the standard deviation, s, of these runs? How does it compare to the standard deviation for the first set of points in the previous problem?

Susan J. LoVerso

Software Engineering

In high school I took the only introductory computer courses that were offered, because my older brother had taken them years earlier. He would bring home pictures that he had generated with the computer and allowed me to color them. So I decided that when I got to high school, I would take computer courses so I could create pictures on the computer, too. Although I wasn't very interested in coloring pictures by the time I got to high school, I did take computer courses, and discovered that my brain seemed to work well in the logical world required by computers.

In college, I made computer science my major, and received a BS/CS from SUNY/Buffalo in 1986, and a MS/CS from SUNY/Buffalo in 1987. But at first, I discovered that I didn't know the first thing about computer science, and there were so many subfields that I didn't know what I wanted to do. After taking courses in graphics, compilers, vision, operating systems, and artificial intelligence, I found two areas interesting. The first was artificial intelligence, a hot field at that time. It seemed really futuristic to program computers that could mimic humans, and I did my master's work in that area.

At the same time, I discovered the world of operating systems, and UNIX in particular. I had to use a new UNIX machine for some of my coursework, and I quickly discovered the wonderful social aspects of the system, particularly the "talk" and "write" programs. The ability to talk to my friends over the computer from miles away—even into the early hours of the morning—hooked me on UNIX. I even met my husband this way. Although my graduate school project was in AI, my job in the department was maintaining the UNIX machines. And the more I learned about UNIX, the more interested I became. I like knowing the nuts and bolts about how things work, and learning the internals of the operating system helped me understand how the machine really worked. In the end, UNIX won as the direction of my career.

In 1987, I joined Encore Computer Corporation as a software engineer. I worked there for four and a half years in the OS group, learning in greater depth many of the subsystems comprising the operating system. During that time, my husband and I decided to start a family. One week after I found out I was pregnant, Encore decided to close the facility where I worked. After finishing up commitments there, now five months pregnant, I started working at Thinking Machines Corporation, where I have been for over two years. I continue to do operating system development for their Connection Machine and to learn more about how systems work.

RAID (Redundant Array of Inexpensive Disks)

Recently, I have been working with parallel I/O devices, specifically disk devices, known as RAID (Redundant Array of Inexpensive Disks). When data is written to one disk, as part of a file, it has a block number associated with it. When you want to retrieve the data, the disk spins to that block and reads the data. With a RAID device, the data for a file is "striped" across many disk drives—some known quantity of data is placed on one drive before going to the next disk drive in the device. The amount of data written to each disk in the device is called the "stripe size", and is defined as:

$$\text{amount written to each disk } * \text{ number of disks in the device}$$

1. Assume disk drives in the device are numbered 0 - (N-1), if there are N disks in the device.
2. Assume that the data in all files start on disk 0.
3. Assume that 16 bytes are written to each drive before moving on to the next disk.
4. Assume that the device contains seven disks.
5. So, bytes 0-15 of a file are on disk 0, bytes 16-31 on disk 1, bytes 32-47 on disk 2, et cetera.

Problem 100.   a. How big is the stripe size?

b. If I create a file and write 4321 bytes to it, the end of file is on which disk?

c. If I begin writing at an offset of 8000 in my file and I write 1500 bytes, on which disks do I begin and end my I/O operation? How many complete stripes do I write and how many partial stripes?

The term "redundant" in RAID implies protection against failure. The simplest form of protection is simply storing a copy of the data from each drive onto another. This is called "mirroring". However, if every disk drive requires another for protection, the costs increase very rapidly. Another technique for protection is called "parity". By taking each byte written to a particular location on a disk and performing an exclusive OR operation, we generate parity. Then, if a single disk fails, we can use the parity and the data from the remaining disk drives to recompute the data that was on the failed disk. The exclusive OR operation (called XOR) is as follows:

$$\text{FALSE OR FALSE } \rightarrow \text{ FALSE}$$

$$\text{TRUE OR FALSE } \rightarrow \text{ TRUE}$$

$$\text{FALSE OR TRUE } \rightarrow \text{ TRUE}$$

$$\text{TRUE OR TRUE } \rightarrow \text{ FALSE.}$$

For these problems assume a device containing three disk drives and one parity drive. Assume that one integer is written to each drive before moving to the next drive.

Problem 101.   What is the parity data if we wish to write out the first nine Fibonacci numbers?

Problem 102. If the following data is on Disk 0, Disk 2, and the Parity drive, compute the data that existed on failed Disk 1.

Disk 0	Disk 1	Disk 2	Parity
21	XX	0	2
34	XX	16	69
55	XX	144	254

On the network in the *misc.jobs.misc* newsgroup, someone made a passing reference to this problem, one they got during an oral exam for their PhD. The answer is so obvious that I was stumped for two days! I was looking for some big, complicated formula, when the answer was there right in my face.

Problem 103. Write down a three-digit number. Then, repeat those same three digits to make a six-digit number (for example, 123123, 941941, et cetera). This six-digit number is evenly divisible by seven. Why?

Stocket Market Investing

Another hobby of mine is investing, specifically in the stock market. Several colleagues and I have started an investment club, to learn how to analyze companies and their stock before deciding whether to invest. I didn't know anything about researching a company prior to joining, and have learned a great deal about how businesses work since then.

Problem 104. One key indicator of a business's health is increased earnings per share (EPS) compared to the EPS of a year ago. My club likes to see at least a 15 percent increase in earnings each year. If Company X had EPS of $1.50 last year, and the first three quarters of this year it had earnings of $.33, $.40, and $.48 respectively, what EPS does Company X need in the fourth quarter to reach a 15 percent increase over last year? If the company had 48 million shares outstanding during those years, what are the annual earnings of the company?

Problem 105. One nice thing a business can do for its shareholders is to buy back its own stock. By decreasing the number of outstanding shares, the remaining shares become more valuable. If Company X had $82.56 million in earnings this year and bought back 8 percent of its 48 million shares, what is the company's EPS for this year? What is the increase over last year?

Eileen Thatcher
Immunology and Microbiology

I attended a three-year junior high school in a small town. Then I went to a very small rural high school for two years. There were only 200 students so class sizes were small. One math class had only six students. Most days were spent working at the blackboard, so we really learned to think on our feet! For as long as I was in school, I studied math—basic algebra, plane and solid geometry, trigonometry, matrix algebra, and analytic geometry.

At UC, San Diego, I earned a BA in Biology. While there, I took three courses of calculus and one course of statistics, and worked in a variety of jobs. As an undergraduate and between research positions, I worked as a bookkeeper. Good math skills helped me get those jobs, and they certainly paid better than the usual student employment. Bookkeeping was okay and I learned useful skills, but it was not nearly as fun as working in a lab.

I also worked in immunology research in cancer research and cellular immunology at the University and at the VA Hospital in San Diego. My bookkeeping skills were valuable when I was working as a lab supervisor, because I was responsible for budget management and developing budget estimates for grant funding. Math was also needed for scientific calculations and statistical analysis. Later, I was co-owner of a business making antibodies for diagnostic and research use, and served as a consultant to other companies involved in antibody production. After years spent watching San Diego County grow and become less rural, I desired something new. So I moved to Humboldt County, which is very different indeed. People have huge redwood trees in their yards—quite a change from the lemon trees of Southern California!

For four years, I was a supervisor of the bacteriology lab complex at Humboldt State University. By then, I had worked a total of nine years and had finally earned enough money to return to school. I wanted to work in immunology and continuing on to graduate school seemed like the perfect way to do that.

So I went to UC, Davis for a PhD in immunology. While working on a research paper, I discovered that although I had a great statistical software package, I could only understand about one-third of the statistical applications and instructions. To teach myself the necessary methods, I spent the next three months studying advanced books on statistics. By this time, it had been years since I had done any formal study in math, and many things I had not used in the meantime had turned to cobwebs in my brain. However, a little review brought the lost material back. Learning newer methods was fun because I had something real and immediate to use as examples—my research.

In addition to graduate study at UC, Davis, I worked in bacteriology research and as a teaching assistant. After graduation, I was a lecturer at UC, Davis, and now I am part of the faculty at Sonoma State University. Currently, I am working on developing assays for monitoring pollution caused by animal waste washed into streams and bays by rainfall run-off. I use dilutions of antibodies to identify the source of animal pollution, and basic statistics to analyze the results of the assays. I work with students who are doing research on a variety of bacterial studies. For example, one student is interested in the effects of oxygen on a group of bacteria that live and grow only in the absence of oxygen. Another student is studying how different temperatures and length of time during shipping can affect the total number of bacteria present in oysters. A third student is busy characterizing bacteria collected from geysers and hot springs. In all cases, dilution techniques are used in their work, either while preparing solutions or as part of their research assays. Statistical analysis helps interpret their results and in designing the next phases of their studies.

In my laboratory classes, I find that most students have trouble with dilutions. After I wrote these problems, I began using them to introduce dilutions and how to work with them. Many students have commented that they wish they had practiced "dilution math" and the concepts behind it much earlier in their education.

The Solution is Dilution

Dilutions are often necessary in chemistry, immunology, and microbiology. Dilution is simply taking a concentrated substance and mixing it in water or another liquid to a desired strength. Serial dilution is a step-wise process of dilution to yield several desired concentrations.

Problem 106. Start with a practical example of diluting chocolate syrup in milk to make chocolate milk.

One tablespoon = 1/2 ounce.

One lb. = 16 oz.

One gal. = 128 oz.

To make chocolate milk, a brand of chocolate directs you to dilute the syrup to one part in sixteen. How many tablespoons of syrup are needed for one eight-ounce glass of milk?

Problem 107. How much chocolate milk can you make with a one-pound can of chocolate syrup?

Concentrations and Dilutions

A basic rule of physics, chemistry, and the universe states that matter is not created from nothing, nor does it disappear or become nothing. However, it can change form. For example, if one teaspoon of salt is added to a glass of water and stirred, the salt appears to disappear. But if you taste the water, you know that the salt is still there. If you boil the salt water, you will find that: 1) the water changed from a liquid to a gas (although it seemed to disappear as steam); and 2) the salt is again a solid. This "conservation of matter" can be applied to matter in liquid form as well. The amount of chocolate in chocolate syrup, a very concentrated form, does not change when added to milk. It just becomes less concentrated. A handy formula based on this conservation of matter is: $C_1 V_1 = C_2 V_2$.

If the amount of a substance, such as chocolate, is equal in two solutions, then the product of [concentration * volume] of one solution ($C_1 V_1$) must equal the product of [concentration * volume] of the other solution ($C_2 V_2$). For example, assume the syrup is at a concentration of 0.5 ounces solid chocolate per ounce of syrup. According to the directions, chocolate milk is 1/16th as concentrated, or 0.03125 ounces solid chocolate per ounce of milk. Problem 106 could be expressed as follows:

C_1 = concentration of chocolate in syrup = 0.5 oz/oz liquid

V_1 = volume of syrup needed

C_2 = concentration of chocolate in milk = 0.03125 oz/oz liquid

V_2 = volume of milk desired = eight oz liquid

Solving for V_1 leads to:

$$V_1 = \frac{C_2 V_2}{C_1}$$

$$= \frac{0.03125 \text{ oz/oz milk} * 8 \text{ oz milk}}{0.5 \text{ oz/oz syrup}}$$

$$= 0.5 \text{ oz (which} = 1 \text{ tablespoon)}$$

Although this is a cumbersome way to solve a simple problem, it is very useful in a wide variety of more complex applications, especially in chemistry and biology.

Okay, the chocolate milk should give you enough energy for more dilutions! Although M stands for "molar," a measure of concentration in chemistry and related fields, you don't need to know chemistry to do this problem. A one-molar (1M) solution contains one gram molecular weight per liter. For example, a 1M solution of sodium chloride (table salt) contains 58 grams of sodium chloride per liter of water, because the molecular weight of sodium chloride is 58. Molecular weight is the sum of the atomic weights of the elements present in the molecule. Atomic weight is based on one atom of an element, such as sodium or chlorine. The atomic weight of sodium is 23 and the atomic weight of chlorine is 35 (23 + 35 = 58). Hydrochloric acid is hydrogen and chlorine as hydrogen chloride in water. The atomic weight of hydrogen is one. Add one to 35 for chlorine to get the molecular weight of hydrogen chloride: 36. Therefore, a 1M solution of hydrochloric acid contains 36 grams of hydrogen chloride per liter of water. It is much easier to work with molar than with gram molecular weights or gram atomic weights.

1000 milliliters = one liter

Problem 108.   Use the following formula for calculating concentrations and dilutions:

$$C_1 V_1 = C_2 V_2$$

[concentration(1) × volume(1) = concentration(2) × volume(2)]. For example, to make one liter of a 0.5 M solution of hydrochloric acid using a 11M concentrated stock acid, how much acid and water do you need?

Dilution Schemes

Speaking of water, what's the best way to determine how many bacteria are found in a sample of water taken downstream from a sewage plant? This is a case where serial dilutions can be helpful. In one method, you set up a series of tubes. Each contains a growth media (bacterial food) with a known amount of sample water added to it. For each tube, pour the media/sample mixture into a flat dish called a petri plate. After a day or two, count the colonies of bacteria that grew in each plate. The simplest way is to first add the sample to dilution tubes. Place the sample into the first tube, mix, then transfer some of the mixture from the first tube to the second tube, mix, and continue the process down the line. In the diagram, four tubes are used. Each contains nine ml water and one ml sample. This is called a ten-fold dilution series.

A ten-fold dilution contains one part of something in ten parts total. Other dilution schemes can be used also. A two-fold dilution contains one part of something in two parts total, a five-fold dilution contains one part of something in five parts total, and so on. Chemists and biologists have different ways of writing dilutions in shorthand. A chemist writes a ten-fold dilution as "1:9", meaning one part plus nine parts. However, a biologist writes a ten-fold dilution as "1:10", meaning one part in ten parts. The difference? A chemist likes to see what needs to be added when mixing, and calculate later; a biologist likes to make calculations easier, and figure out what is needed later, when mixing. Both ways work. The trick is knowing who wrote the instructions before trying to follow them yourself!

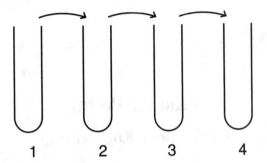

1   2   3   4

Tube #1: 9ml water + 1ml sample
Tube #2: 9ml water + 1ml from tube #1
Tube #3: 9ml water + 1 ml from tube #2
Tube #4: 9ml water + 1ml from tube #3

FIGURE 47
Sample tubes

Problem 109. What is the final dilution in tube #4? In other words, what fraction of initial concentration of sample [added to tube #1] is in tube #4? If you add sample directly to tube #4, how much initial sample will you add to yield the same concentration achieved by the serial dilution method?

Problem 110. Five milliliters of each tube is added to growth media and poured into a petri plate, and then allowed to grow bacterial colonies. Each colony represents one bacterium in the sample. A reasonable range for counting colonies on plates is 30 to 300. The table below gives the results of counts for a set of four plates. What is the bacterial count per milliliter of initial sample?

Tube # :	1	2	3	4
Dilution:	10^{-1}	10^{-2}	10^{-3}	10^{-4}
Count :	TMTC*	260	21	3

*TMTC—too many to count

Problem 111. Following the pattern in tube #4 above, set up a two-fold dilution series. Show how much sample and how much diluent (dilution water) is put into each tube. What is the final dilution in tube #4?

Julie A. Pollitt
Mechanical Engineering

I have been interested in math for as long as I can remember, especially the challenge of solving a problem. In my junior year of high school, I won Math Student of the Year. My teachers were pushing me to choose a career as a math teacher, but I wanted to work for the National Aeronautics and Space Administration, better known as NASA. A college recruiter suggested aerospace engineering, and informed me that engineering requires plenty of math. So I enrolled in college with the intention of becoming an aerospace engineer, but when I spoke to several senior engineering students, they recommended mechanical engineering instead, because it is a broader field. Aerospace is very similar to mechanical engineering, but concentrates on space and aerodynamics classes. In mechanical engineering, there is no specific concentration, so you can take more general classes. Needless to say, I took a lot of extra math classes. In 1988, I graduated cum laude with a Bachelor of Science in Mechanical Engineering from the University of Connecticut. Then, I left for California and a job with NASA at Ames Research Center.

There, I work as a mechanical systems design engineer. My group is a support organization for the rest of Ames. That means that scientists—biologists, physiologists, psychologists, chemists, physicists, aerodynamisists, etc.—ask my group to design the mechanical apparatus needed for their research. We also analyze existing mechanical systems and improve them. These apparatus can be anything from an airplane, to enclosed chambers used to grow plants in space, to the Space Shuttle. My job is every bit as exciting as I envisioned it would be when I was in high school.

One design I was responsible for was a cooling system for the Crop Growth Research Chamber (CGRC). The CGRC is a ground-based chamber that is the precursor of an enclosed chamber to grow plants in space. The CGRC has its own environment. The composition of the air and plant nutrients, along with the ambient temperature and pressure, can be varied to study the effect of these variables on plant growth. Our engineering group worked for almost a year designing and building the entire earth-bound chamber. So far, several different plant varieties have been grown in the chamber with favorable results, i.e., a large yield-per-growth area. As part of my job, I was required to get a master's degree. So in March of 1991, I received a Master of Science in Mechanical Engineering from Stanford University. Until it is time to pursue a doctorate degree and teach mechanical engineering at a university, I plan to stay with NASA.

F-18 Hornet Test

For a future design application, a group of NASA engineers need to know how much extra air can be removed from the jet engines on an F-18 Hornet fighter/attack aircraft. The engines, in the rear of the plane, take in air that is mixed with fuel and then ignited. Some of the air is removed from ports on each engine via ducting along the back of the aircraft, and brought to the front. This air provides heat for the pilot, keeps the windshield free of ice, and—after passing through several heat exchangers—cools the electronic equipment in the nose of the airplane.

To remove the excess air, the engineers want to try installing a six-inch duct that branches off the existing duct. The six-inch duct must carry the air—which is at 1000°F—away from the plane before allowing it to exit into the atmosphere. It would be difficult to place the extra ducting within the aircraft and fly, so a ground test (with the airplane tied to the ground), is performed instead. The duct is suspended above the wing and parallel to the ground at a height of 118 inches (measuring to the centerline of the duct). The duct descends at a 45° angle with the ground before continuing along the ground away from the plane.

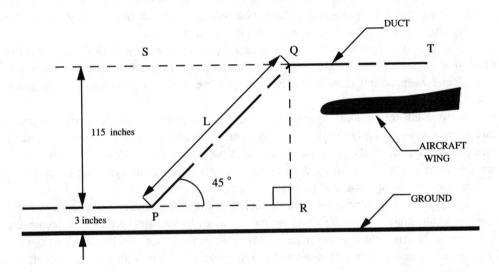

FIGURE 48

Problem 112. How long (L) is the piece that goes from the duct above the wing to the duct on the ground?

Helen Townsend-Beteet

HMO Pharmacy Practice and Management

By the time I reached my senior year in high school, I was certain I would become a clinical psychologist. My mother was pursuing a masters in guidance and counseling, so there was a family influence. Fortunately, my high school classes included mathematics and the sciences.

I began my studies in psychology at the University of Kansas. My classes were enjoyable but I became concerned that the field might be overcrowded and offer little real career growth. A minority affairs counselor suggested that I try a field called psychopharmacology. It involves studying drugs to see how they affect receptors in the brain.

So at the end of my sophomore year, I switched to a pre-pharmacy curriculum. That meant I had to attend school for an additional year to take the courses required for pharmacy school. I have never regretted this extra year because it gave me more job options and a comfortable salary.

My work history includes two and a half years of retail pharmacy and ten years of managed care pharmacy with a health maintenance organization. Pharmacists do far more than just fill prescriptions. I particularly enjoy interacting with physicians and other health professionals, and educating the public on medication matters. As a supervisor of a growing department, I always have reports, analysis, and processes to improve.

Today, I am married with two small children, and am also pursuing a masters in health services administration. I've already taken more classes in statistics, economics, accounting, and finance. In this field, one never gets away from mathematics! An advanced degree will help me stay in tune with the economic and quality issues associated with our country's attempt at healthcare reform.

Preparing Medication

Not all drugs are commercially available, and pharmacists are often asked to prepare medications in forms that are not manufactured. In medical practice, a physician may choose to prescribe a medication to be created by the pharmacist, or the pharmacist must research how to prepare a product in a reduced strength. For example, many pediatric medications are first formulated in adult doses.

Problem 113. Pharmacist Jan Sharp has received a prescription for a drug known as Trispin.* Trispin is available only as a 10mg tablet. Manufacturer's studies show that Trispin can be used in infants at a dose of between 1mg to 3mg, four times a day. After reviewing the drug's solubility, Jan knows she can mix the crushed tablets with cherry syrup to form a solution that will be stable for one month. The prescription is for 3mg to be given four times a day and instructs Jan to dispense a one-month supply. How many tablets of Trispin and how much cherry syrup will be required for a 3mg-per-teaspoonful dose? (One teaspoonful = five ml.)

Problem 114. How will the answer to Problem 113 change if the dose is reduced to 1mg/2.5ml four times a day for 15 days?

Problem 115. Many products are dosed by weight. For example, the Trispin dose is .5mg/ kilogram for infants. What dose could an 18-lb infant take? Round the answer to two significant figures. (One kg = 2.2 lbs.)

Problem 116. Besides weight, a particular dose of a drug may be based on a person's body surface area. The usual adult body surface area is expressed as 1.73 square meters (1.73 m^2). If Trispin is also dosed 30mg to 40mg per square meter, what would the maximum dose be for a 1.9 m^2 adult?

Diluting a Solution

Sometimes the pharmacist has exactly what is prescribed in stock, but there are times when a particular strength must be compounded. The pharmacist may need to dilute a concentrated solution to create a product of lesser strength. Concentrations can be thought of as percentages. For example: Ten grams of sodium chloride (NaCl) in 100 ml of distilled water is a 10 percent solution. Common pharmacy concentrations are: grams per gram (weight/weight), grams per milliliter (weight/volume), and milliliter per milliliter (volume/volume). Concentrations are expressed as:

$$\frac{\text{quantity of solute}}{\text{quantity of the solution}}.$$

In the example of sodium chloride, concentration is expressed as:

$$\frac{10 \text{ g NaCl}}{100 \text{ ml of distilled water}} = \frac{1}{10} = 10 \text{ percent solution.}$$

---

* Trispin is not an actual product.

Problem 117.   Pharmacist Jan Sharp receives a prescription for a 20 percent solution of Compound A to be used for contact dermatitis. When she inspects the shelves, she discovers there is only a 50 percent stock solution of Compound A on hand. She decides to prepare the requested strength by diluting the stock solution with distilled water. How much of the 50 percent solution of Compound A and how much distilled water will be required to meet the required 20 percent strength? Stock on hand is listed below, along with the prescription:

One pint (240 ml) Compound A, 50 percent solution
One pint (240 ml) distilled water

Name: I. M. Itching
Compound A 20 percent: 120 ml
Sig: Apply to affected areas BID (twice daily) as directed
Brand exchange: Permissible
Refill: Four times

Problem 118.   The contact dermatitis (rash) has improved greatly, but the doctor would like the patient to continue on a reduced strength of 10 percent until the rash clears. Jan needs to make 120 ml of a 10 percent solution. She still has the 50 percent stock solution of compound A. How much stock solution and distilled water will she need this time?

Problem 119.   The patient's contact dermatitis turns out to be eczema. The doctor wants Jan to prepare the following:

LCD solution five percent and
Aspirin powder five percent
In 60 ml of .1 percent Tiamcinolone
How much of each ingredient will be needed?

Jane D. Kivlin

Ophthalmology

I have always enjoyed mysteries. Figuring out "whodunit," tracking the culprit down, and then proving it fascinates me. I approach mathematics problems in the same way. How can I take the information and get the answer I am looking for? This way of looking at mathematics makes even story problems interesting! And mathematics is so much more precise than the psychology of people and their temptations, I can get an exact answer, usually without reading 250 pages.

Despite my liking for mystery novels, I am very practical in nature. The application of calculus to practical matters interests me. One problem I specifically remember puzzling over is, "What is the most economical measurement for a conical paper cup?" What is the best way to maximize the volume the paper cup can hold while minimizing the surface area?

Another interesting problem was a question about whether a person could stay at the tip of the shadow of a Greek philosopher's walking stick as the sun started to set. In the early afternoon, the person could only inch forward. But the shadow would grow astronomically in length just before the sun set, making it impossible to run as fast as the tip of the shadow was going.

While the precision of problem-solving in physics and chemistry appeals to me, my more practical side chose medicine. Nothing in medical school was as interesting as ophthalmology (the study of the eye), a course I took as an easy summer class! The eye is such a small part of the body, but it is so complex. It can detect a huge range of light levels and measure very small distances. We not only have fine reading vision, our eyes also have the ability to provide peripheral (or side) vision, color, dark and light contrast, movement, and distance. We can even measure how surgery improves the eye's ability to function much more precisely than we can measure the affect of surgery elsewhere on the body.

In my research, I specialize in the study of how the eyes can be used together to achieve a sense of three-dimensional vision. I have subspecialized in the area of strabismus. This relates to the way that both eyes need to be pointing at the same object to send a coordinated image of the outside world to the brain. When the eyes are not straight, the person has strabismus. Most commonly, the eyes turn in toward the nose, but they can be deviated in an outward direction, up or down, even rotated relative to each other. In my surgery, I can either strengthen or weaken the eye muscles to bring the eyes into a better position. I enjoy doing surgery, because in a few short hours, I can make a big improvement in how well my patient sees and, as a secondary benefit, what the patient looks like.

With a full-time job at a medical school, a husband, and a daughter, I read mysteries by listening to books on tape in the car or while wearing headphones while I do household tasks.

Having a professional career means that I work hard, but my schedule is my own. My husband and I try to take turns attending our daughter's school functions, but planning ahead is essential. Sometimes when my daughter is ill or on vacation, we have to make quick arrangements for someone to be with her. With two incomes, we can pay for help when necessary, and that certainly helps us to cope.

Double Vision

After a head injury, a patient complains of double vision. The images are very close but she cannot quite put them together. With a small amount of prism in her eyeglasses, the images can be shifted enough to allow her to see one image with depth perception.

Different prisms can shift an image different amounts, depending on the angle between the front edge of the glass (where the light enters), and the back edge (where the light exits at a different angle on its way to the eye). A very large angle shifts the light more than a very small angle. The amount the light shifts is called the power of the prism, and the unit used to measure the power of a prism is called a prism diopter.

This patient needs 12 prism diopters horizontally and five prism diopters vertically. A prism that corrects both the horizontal and vertical requirements can be set at an oblique angle in the left lens of her glasses.

Problem 120.   a. Using a vector analysis, the patient's requirements can be diagrammed as:

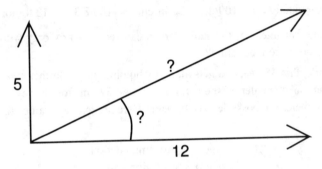

FIGURE 49
Vector analysis of prism

What power of prism is needed in the lens? At what angle from horizontal does she need the prism?

b. This will make the left lens heavy compared to the right lens, so the prism should be split between the two lenses. What power of prism is needed in each lens, and at what angle?

Time for Reading Glasses

Everyone over 45 years of age needs glasses for reading. The lens inside the eye becomes stiffer with age and can no longer focus on near objects. Actually, the focusing ability of the lens starts to decrease in childhood. A six year-old can see near objects more clearly than a high school student can see the objects at the same distance.

Problem 121.   a. Focusing ability is measured in diopters. Knowing how close a person can bring an object and still see it clearly, the doctor can calculate how much focusing ability or accommodation the person has. For example: a 15 year-old girl can bring a book to within 8.3

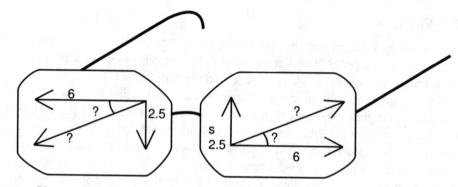

FIGURE 50
Vectors for two lenses

centimeters of her eyes and still see it. That distance is called her near point.

$$\text{Accommodation} = 100/\text{distance in cm} = 100/8.3 = 12.0 \text{ diopters.}$$

However, when she becomes 45, the same person can see it clearly only within 28.5 cm from her eyes. What is her accommodation?

b. For a while, this 45-year-old gets along by holding papers farther away from her, but of course, everything looks smaller. If she can hold a book 57 cm from her eyes, at what age will this become a completely impossible way to read? Here are average values for accommodation according to age:

Age (Years)	Accommodation (Diopters)
45	3.5
50	2.5
55	1.75
60	1.00
65	0.50

c. Most people cope with decreasing accommodation by wearing reading glasses. The dioptric power of the glasses is added to the person's accommodative ability both in life and in calculations. If a 60-year-old wants to read at a 33-cm distance, what power reading glasses does the person need?

Eye Drop Side Effects

Many different kinds of eye drops are used for the treatment of eye diseases. They are absorbed through the clear cornea in the front of the eye, but some of the drop comes out of the eye, and some will go down the tear duct. The tear drainage system starts out as small holes in the upper and lower lids near the inner corner of the eye. Small tubes from these holes travel under the skin, and meet at the side of the nose. A single tube drains to the inside of the nose. Tears normally drain slowly into the back of the nose, where they run down into the throat and are swallowed. When there is a flood of tears, the nose becomes full.

As much as ten percent of an eye drop can eventually be swallowed. Thus, one eye drop can affect the rest of the body.

Problem 122.   a.   A baby is given one drop of one percent Atropine solution to each eye. A one percent solution has 1000 milligrams of medication in 100 milliliters of solution. If there are 20 drops in one milliliter, how much Atropine did the baby receive?

b. How much of the Atropine can be expected to go down the tear ducts and be swallowed, if we assume that ten percent of it will be swallowed?

c.   The baby weighs seven kilograms. The dose of Atropine given by injection before surgery is 0.01 mg per kg body weight. How does this dose compare to that calculated from the eye drops?

Sharon G. Lum

Electrical Engineering

The turning point of my lifelong mathematical awareness was in the sixth grade, when math became fun and interesting. My career plan was laid out in the ninth grade when I decided to become a math teacher, at that time the only job option I knew for people who liked math. Since becoming a math teacher meant taking a lot of math classes and going to college, my high school courses consisted of four years of math, science, and English, two years of Spanish, and of course the required social studies and P.E. classes, plus some music classes. This prepared me for almost any college while providing a well-rounded education.

Attending the University of California at Davis to pursue a mathematics degree, I plunged into calculus, chemistry, and physics. But by the end of my freshman year, mathematics, especially calculus, became too abstract for me. At this same time, I enrolled in an introduction to computer programming class for engineering students and enjoyed it. To my surprise, the engineering students in this computer programming class were also in my math, chemistry, and physics classes. So I took more engineering classes and discovered they were fun. Engineering, especially electrical engineering, uses even more applied math than math majors use. Now I could apply my math and science "tools" to something that felt closer to real life.

After receiving a Bachelor of Science in Electrical Engineering, I went to work for IBM, and have been there ever since. In the meantime, I've also received a Master of Science in Electrical Engineering. Throughout my career at IBM, I have spent most of my time working on and designing in-house automated test equipment for large system hard disk drives. Working with other people in my team, I design circuits, draft test specifications, analyze data, write software programs, run experiments, investigate problems, write documentation, give presentations, learn about new technologies, and travel.

To me, the exciting thing about being an engineer is working with a team and watching projects progress from start to finish: beginning with the initial design, deciding how to tackle the part of the project I am assigned to design. Then following the project through the implementation phase, including building and debugging my design and making sure not only that it works, but also that it works along with everyone else's designs. The last phase may take a couple of days or a couple of months. Even if my design works, it can always be improved, by running experiments, analyzing data, and more debugging. Finally, it's perfect. Time to sit back, relax, and reflect on a job well done (at least for a little while)!

Disk Sweep Head Scan Calculation

Hard disk drives, which are used in personal computers as well as large computer systems, are tested prior to customer use. Checking the disk surfaces for mechanical defects such as scratches and particles is one of these tests. To improve disk yield, a "sweep scan" pass is used prior to the actual test pass. The sweep scan pass is performed by moving a test head slowly across the disk surface from the outside diameter to the inside diameter. Since test time is also a valuable commodity, it is important to optimize the speed of the sweep scan.

Problem 123. Given these parameters:
  Head length: $L_h = 0.16$ in.
  Disk rotational speed: $v = 3600$ rpm (revolutions per minute)
  Disk radius, maximum: $R_{max} = 6.75$ in.
  Find the optimum sweep scan speed so the head "edge" "scrapes" the entire disk surface (see Figure 51). If the scan speed is too fast, the head edge will not scrape the entire disk surface (see Figure 52). If the scan speed is too slow, test time will be wasted.

Voltage Divider Calculation

A particle counter is used to monitor the cleanliness of a hard disk drive test system. If the particle count goes above a certain level, possibly due to a head-disk interaction (HDI or "head

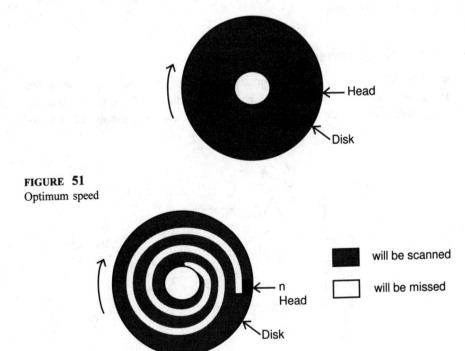

FIGURE 51
Optimum speed

FIGURE 52
Speed too fast

crash"), an alarm signal is sent to the computer that controls the hard disk drive test system, immediately stopping the test.

The alarm signal from the particle counter triggers an audible alarm, using a higher voltage than the signal used by computer systems. The alarm-signal voltage from the particle counter must be reduced to make it usable by the computer system. A couple of simple laws from electronics and a textbook circuit are all that is needed to accomplish this task.

Basic electrical laws

Ohm's Law. Current is proportional to potential difference, or expressed as an equation:

$$V = I * R.$$

V = voltage, such as the voltage (or electrical potential difference) between the terminals of a battery or the slots of a household outlet, measured in units of volts

I = current, a measure of the motion of electrical charges, measured in units of amperes(commonly abbreviated as amps)

R = resistance, the constant of proportionality in this equation and a measure of the amount of electricity that a material can conduct, measured in units of ohms (W)

Kirchoff's Voltage Law (KVL). The sum of the voltages around any closed loop in a circuit is zero (conservation of energy).

Problem 124. Given the parameters and circuit in Figure 53, find the resistance R_1 and R_2. Input voltage: $V_{in} = 15$ volts (from particle counter alarm)

Output voltage desired: $V_{out} = 4$ volts (typical "on" voltage used by computer systems)

Current desired: $I = 0.010$ amps

Computer Numbering Representations

In school, hard disk drives are used with computer systems to store data such as student enrollments and grades. Since computer systems can only handle data in terms of ones and zeroes (or binary), data stored on the hard disk must also be stored in "patterns" of ones and

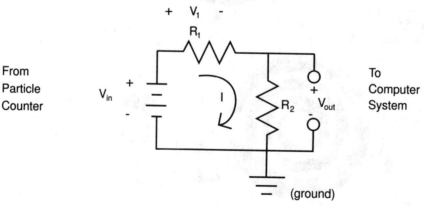

FIGURE 53
Voltage divider circuit

zeroes. A number is an abstract quantity that is represented by symbols. For example, the quantity $*****$ can be expressed as 5 in a decimal, V as a Roman numeral, or 101 as a binary (explained below).

The 5, V, and 101 in the above example are really "names" of numbers. A numbering system provides a format to name numbers with a given set of symbols. People count by tens using the decimal system, possibly because we have ten fingers. The prefix "deci" means ten. Computers count by twos using the binary system, since they can work with only two conditions. The prefix "bi" means two.

Some computers can perform arithmetic operations on large binary numbers at a rate of a million operations per second. For people, these binary numbers are not convenient, so two shorthand numbering schemes are used: octal (the prefix "octa" means eight), and hexadecimal (the prefix "hexa" means six and "deci" means ten).

For example, the binary number 111110101001 is hard to remember, but represented in octal as 7651, or in hexadecimal as FA9, it is much easier to work with. The four most common number systems used by people who work with computers are:

Decimal	base 10	symbols 0, 1, 2, 3, 4, 5, 6, 7, 8, 9
Binary	base 2	symbols 0, 1
Octal	base 8	symbols 0, 1, 2, 3, 4, 5, 6, 7
Hexadecimal	base 16	symbols 0, 1, 2, 3, 4, 5, 6, 7, 8, 9, A, B, C, D, E, F

Hexadecimal Conversion Table:

Decimal	Hexadecimal
10	A
11	B
12	C
13	D
14	E
15	F

Notice that the lowest symbol is always zero and the highest symbol is always "base" one.

To convert a decimal number $N_{\text{base 10}}$ to binary, octal, or hexadecimal, perform decimal divisions until they cannot be done any more, keeping track of the remainders. Divide N by b:

$$
\begin{array}{rl}
& 0 \qquad\qquad \text{remainder } r3 \\
\text{base } b \,\big|\, & \overline{N2} \qquad\quad \text{remainder } r2 \\
\text{base } b \,\big|\, & \overline{N1} \qquad\quad \text{remainder } r1 \\
\text{base } b \,\big|\, & \overline{N_{\text{base 10}}}
\end{array}
$$

Perform the division from the bottom up, then read the remainders from the top down to get the answer:

$$\text{decimal number } N_{\text{base 10}} = r3 \; r2 \; r1 \text{ in base } b.$$

For example, convert 13 base 10 to binary (base 2):

$$
\begin{array}{rl}
0 & r = 1 \\
\hline
2\,\lfloor 1 & r = 1 \\
\hline
2\,\lfloor 3 & r = 0 \\
\hline
2\,\lfloor 6 & r = 1 \\
\hline
2\,\lfloor 13 &
\end{array}
$$

Thus $13_{10} = 1101_{\text{base }2}$.

Another example: convert 2710 to hexadecimal (base 16):

$$
\begin{array}{rl}
0 & r = 1 \\
\hline
16\,\lfloor 1 & r = 11,\ \text{or B} \\
\hline
16\,\lfloor 27 &
\end{array}
$$

Thus $27_{10} = 1B_{16}$. (You can use either "b" or "B" to represent 11 in hexadecimal.)

In the decimal system, the numbers are represented as sums of ones (10^0), tens (10^1), hundreds (10^2), thousands (10^3), et cetera. Similarly, for the binary system, numbers are represented as sums of ones (2^0), twos (2^1), fours (2^2), eights (2^3), et cetera. So for any number system:

Let $b = $ base

Let $x2\ x1\ x0 = $ number representation in base b

Then to convert to decimal:

$$(x2 * b^2) + (x1 * b^1) + (x0 * b^0) = N_{\text{base }10}$$

For example, convert 1010_2 to decimal:

$$
\begin{aligned}
1010_{\text{base }2} &= (1 * 2^3) + (0 * 2^2) + (1 * 2^1) + (0 * 2^0) \\
&= (1 * 8) + (0 * 4) + (1 * 2) + (0 * 1) \\
&= 8 + 0 + 2 + 0 \\
&= 10_{\text{base }10}.
\end{aligned}
$$

And convert $47_{\text{base }8}$ to decimal:

$$
\begin{aligned}
47_{\text{base }8} &= (4 * 8^1) + (7 * 8^0) \\
&= (4 * 8) + (7 * 1) \\
&= 32 + 7 \\
&= 39_{\text{base }10}.
\end{aligned}
$$

Binary addition is performed the same way as decimal addition, except that only binary, 0 and 1, symbols can be used. Using some examples for illustration:

$$
\begin{array}{ccccc}
0 & 1 & 10 & 11 & 10 \\
\underline{+\,1} & \underline{+\,1} & \underline{+\,1} & \underline{+\,1} & \underline{+\,10} \\
1 & 10 & 11 & 100 & 100
\end{array}
$$

The purpose of these problems is to acquaint you with the number systems commonly used by computer systems, and to practice adding numbers the way a computer does it.

Problem 125.   a.   Convert the following numbers from decimal (base 10) to binary (base 2):

$$8_{\text{base }10} = \underline{\quad}_{\text{base }2} \qquad 3 = \underline{\quad} \qquad 12 = \underline{\quad} \qquad 5 = \underline{\quad}.$$

b.   Convert the following numbers from decimal to octal (base 8):

$$8_{\text{base }10} = \underline{\quad}_{\text{base }8} \qquad 3 = \underline{\quad} \qquad 12 = \underline{\quad} \qquad 25 = \underline{\quad}.$$

c.   Convert the following numbers from decimal to hexadecimal (base 16):

$$8_{\text{base }10} = \underline{\quad}_{\text{base }16} \qquad 3 = \underline{\quad} \qquad 12 = \underline{\quad} \qquad 20 = \underline{\quad}.$$

d.   Convert the following numbers from the given bases to decimal:

$$1101_{\text{base }2} = \underline{\quad}_{\text{base }10} \qquad 10000_{\text{base }2} = \underline{\quad}$$

$$17_{\text{base }8} = \underline{\quad} \qquad 20_{\text{base }8} = \underline{\quad}$$

$$E_{\text{base }16} = \underline{\quad} \qquad 2D_{\text{base }16} = \underline{\quad}$$

e.   Add the following binary numbers. Check the results by performing the addition in decimal. Remember that only ones and zeroes are allowed in binary arithmetic.

1010	1010	101	1111	1011
+ 100	+ 10	+ 1101	+ 11	+ 110

Beth MacConnell
Fish Pathology

I grew up in the suburbs of New York City, and moved West when I graduated from high school. Science, biology in particular, was interesting so I also took the required math courses algebra, trigonometry, and geometry. A college degree in biology required even more math. Although it was never easy for me, I am very glad I didn't give up on math. Little did I know that a career in biology would require putting my math skills to use on a regular basis.

After graduating from Colorado State University with a BS degree in Wildlife Biology, I worked for one year as a research assistant with the Colorado Division of Wildlife. I counted deer in remote areas for population assessments. To keep the data as unbiased as possible, the paths we took through the mountains were determined by a computer. This meant having to pick my way through boulder fields instead of taking the easy way around.

I moved to Bozeman, Montana and worked as a biological technician on the Interagency Grizzly Bear Study Team for four years. At times, this was very exciting work, especially when I got up close and personal with the grizzly bears we trapped and radio-collared. My summers were spent in a small plane, tracking those bears. Flying at tree-top level and hanging out of the plane to take photographs was often heart-stopping. One of my winter assignments was to use the mapped sightings, a planimeter, and my math skills to determine the size of the area each bear used. During this time, I got married and had a baby. Getting up at 4:00 am to go flying and often being gone overnight made it difficult to balance family and work.

After working as a fisheries biologist for a few years and learning about fish pathology, I pursued a Masters degree in veterinary science at Montana State University. For the past seven years, I have been working as a fish pathologist for the U.S. Fish and Wildlife Service. Preserved fish from a wide variety of places including hatcheries and rivers in the United States, Canada, and even South America are sent to me for disease diagnosis. I use mostly histological techniques, and recommend treatments. My math skills are used every day, from determining water flows in a stream, making chemical solutions in the lab, or administering a therapeutic treatment to fish in a raceway. Incorrect calculations can be disastrous, especially for the fish.

My son is a senior in high school now. I travel more every year to collect fish from contaminated rivers, teach fish health and diseases, fish histopathology, and attend meetings.

Disease Treatment

Aquaculture (growing fish) is a fast-growing industry worldwide. Fish raised intensively often require treatment for one disease or another during a rearing season. The drugs and chemicals that are used to control infectious organisms, such as bacteria or parasites, can be toxic to fish if concentrations are too high. Treatment calculation is critical and should be double-checked before being administered. In human or veterinary medicine, patients are treated on an individual basis under carefully controlled conditions. Fish are treated as groups—often hundreds of thousands of individual fish.

There are several methods for treating fish: dip, constant-flow, prolonged bath, and indefinite bath. Constant-flow treatment is most commonly used by fish hatcheries in raceways, tanks, or troughs where it is impractical or impossible to shut off the inflowing water. The volume of water flowing into the unit must be determined so a solution of the chemical treatment can be metered into the inflowing water to obtain the desired concentration.

ppm = parts per million

Calculation for constant-flow treatments:
(rate * time * concentration * correction)/strength = weight
rate = flow rate
time = treatment time
concentration = final concentration
correction = correction factor. Convert volume to weight in grams (or milliliters) per
gallon = 0.0038 correction
strength = chemical strength as an decimal fraction
weight = weight of chemical needed

Problem 126. A trough with a water flow of six gallons per minute is to receive a one-hour (60 minute) constant-flow treatment of chemical B (100 percent active strength) at a concentration of five ppm. How many grams of chemical B must be dispensed to maintain the treatment concentrations?

Problem 127. A raceway with a water flow of 300 gallons per minute is to receive a one-hour (60 minute) constant-flow treatment of chemical C (80 percent active strength) at a concentration of two ppm. How many milliliters of chemical C are needed?

Gas Supersaturation

Nitrogen and oxygen are the most abundant gases dissolved in water. Gas supersaturation of water can occur when air is introduced under high pressure (e.g., dam spillways, deep wells). All fish are susceptible to gas bubble disease, which can be caused by any supersaturated gas in the water, but is usually due to excess nitrogen. Reduced pressure on the gas, or increased temperature, can bring nitrogen out of solution to form bubbles in the bloodstream, similar to the "bends" in human scuba divers. Tolerances to supersaturation vary among fish species, but any saturation over 100 percent can pose a threat to fish health. Changes in atmosphere, pressure, or temperature will cause dissolved gases in water to change. To determine percent saturation and nitrogen, the water barometric pressure, temperature, dissolved oxygen, and gas pressure must be measured. The Bunsen coefficient and vapor pressure must also be determined.

In fisheries biology, it is important to know the percent saturation and nitrogen levels of the water. Continuous monitoring is necessary in rivers below dams and hatcheries that use well water.

Pressure is measured in millimeters of mercury, Hg (mmHg).

Total gas pressure = DP (mmHg)
Barometric pressure = BP (mmHg)
Temperature = T (°C)
Dissolved oxygen = DO (mg/Liter)
Bunsen coefficient of oxygen= BC (L real gas / L atmosphere)
Vapor pressure = VP (mmHg)
Correction factor 1 = CF1 = 0.532
Correction factor 2 = CF2 = 0.790

$$\%\text{Saturation} = \frac{BP + DP}{BP} * 100$$

$$\%\text{Nitrogen} = \frac{(BP + DP)\left[\frac{DO}{BC}(CF1)\right] - VP}{(BP - VP)CF2} * 100$$

Anglers have reported catching fish on the Best River, just downstream from the Big Dam, that have air bubbles in their skin, fins, and mouths. Fungus has infected the wounds created by the broken air bubbles—sounds like gas bubble disease. Determine the level of gas supersaturation and nitrogen in the water, and decide whether reduced spill over the dam is needed to reduce gas supersaturation.

Problem 128.　a. Barometric pressure is 587, temperature is 14°C, dissolved oxygen is 7.2, and total gas pressure is 38. Tables say that at 14°C, vapor pressure is 11.98 and Bunsen coefficient is 0.035. What is the level of gas saturation in the water from the Best River? What is the level of nitrogen in the water?

　b. If the temperature dropped to 10°C, DP to 35, DO rose to 10, and BP was the same (BC = .038, VP = 9.2), would nitrogen gas be a problem? Calculate both the level of gas saturation and nitrogen in the water.

Barbara Swetman

Computer Science and Computer Graphics

As early as I can remember, I enjoyed math in school. It may be difficult for some to believe, but I actually thought that those long homework assignments were fun—though definitely not a breeze!

My high school courses consisted of algebra, geometry, trigonometry and pre-calculus. In 1983, I graduated from St. John's University in Jamaica, New York with a BS in Mathematics. A common question people asked as I went through college was, "So what are you planning to do with your math degree—teach?" Well, I knew that teaching was not for me, so I sought a secondary area of concentration. At that time, St. John's math curriculum offered majors in actuarial science (probability and statistics within the insurance field) and computer science. Since actuarial science required a long-term commitment through ten grueling certification exams, I decided to go with the latest and greatest high tech field. I haven't regretted it since!

In college, I took these math courses: Calculus I, II, III, & IV, Linear Algebra, Honors Seminar in Mathematics (i.e., History of Math), Differential Equations, Complex Variables, and Elementary Number Theory. My computer science courses consisted of: Introduction to Computer Graphics, Introduction to Computing, Fundamentals of Computer Science, Digital Circuits & Computer Organization, Programming Languages, Applied Computer Science, and Systems Software I. In addition, I took Operating Systems and Database Concepts as post-graduate computer science courses.

For six years, I worked at Grumman Space Systems in Long Island, New York, primarily doing software development for one of President Reagan's "Star Wars"-funded programs. My job involved image processing, scientific analysis, and serving as assistant manager of an engineering computer facility. Then, for a little over a year, I worked at Pixar, a California graphics company specializing in high-end 3-D graphics and photorealistic rendering. I performed technical customer support for their 3-D image processing and rendering software, as well as some quality assurance. For the past year, I've been providing customer support for Auto-trol Corporation, a CAD-CAM software company in Denver.

Binary conversions

The computer industry has agreed upon one universal way to store alphanumeric characters (a, b, ..., A, B, ..., 1, 2...) in a computer. This standard representation, known as ASCII, simply associates each character with its own unique numeric value within the range 0 to 255. For example, the letter "A" has been assigned the decimal (base 10) value 65. Since the computer can understand only zeroes and ones ("on" and "off"), these ASCII values are really stored in the computer as binary (base 2) numbers. Each binary-coded character is eight bits (or one byte) of data. So "A" actually appears in the computer as:

$$65_{10} = (1 * 2^6) + (1 * 2^0)$$
$$= (0 * 2^7) + (1 * 2^6) + (0 * 2^5) + (0 * 2^4) + (0 * 2^3) + (0 * 2^2) + (0 * 2^1) + (1 * 2^0)$$
$$= 0100\ 0001_2$$

Binary is commonly written in sets of four digits, separated by a space. Even for people who work with binary numbers, it is difficult to recognize instantly the value of a long string of zeroes and ones. But after working with binary for a while, it becomes easy to recognize the value of any string of four binary digits, since there are only 16 possibilities:

$$0000,\ 0001,\ 0010,\ 0011,\ 0100,\ 0101,\ 0110,\ 0111,$$

$$1000,\ 1001,\ 1010,\ 1011,\ 1100,\ 1101,\ 1110,\ 1111$$

Those are the numbers zero to 15, written in binary. Long binary numbers are represented as strings of these four-digit binary numbers. Since the conversion to decimal can still be somewhat tedious, ASCII code is often referred to in an intermediate representation known as hexadecimal (base 16) code. It provides a much easier method of converting between decimal and binary, because each hexadecimal digit consists of four binary digits.

The binary numbers above are the same as these hexadecimal numbers:

$$0, 1, 2, 3, 4, 5, 6, 7, 8, 9, A, B, C, D, E, F$$

Be careful not to confuse the hexidecimal numbers A through F with ASCII characters. The hexidecimal number A is equal to the decimal number ten or the binary number 1010. The letter "A" is represented in the ASCII standard by the decimal number 65.

As stated above, "A" actually appears in the computer as:

$$65_{10} = (1 * 2^6) + (1 * 2^0) = 0100\ 0001_2$$

but writing down $(1 * 2^6)$ means you have to carefully place the one in the sixth place from the right. This takes too long, especially for powers greater than six. It is easier to use hexadecimal:

$$\text{"A"} = 65_{10} = (4 * 16^1) + (1 * 16^0) = 41_{16}$$

To find out how the computer would store 41_{16} is an easy translation. Since $4 = 0100$ and $1 = 0001$, then the binary representation of 41_{16} is $0100\ 0001_2$, the same answer as when we translated directly to binary.

Since each hexadecimal digit represents four binary digits, writing numbers in hex takes four times fewer digits than writing in binary. That makes it significantly more concise and easier to understand.

Decimal is not as useful as hex when dealing with numbers that are ultimately represented in binary, because there is no clean ratio between decimal and binary. There is no way to visually translate 65 into $0100\ 0001_2$ without doing the calculation.

Problem 129. Given that the ASCII representation of the upper-case alphabet ranges from 65 to 90 (decimal) and lower-case ranges from 97 to 122, how is the word "Math" represented in binary using standard ASCII? (The translation will take fewer steps if you translate each letter into hex and then convert those hex numbers into binary.)

Problem 130. What word do these hexidecimal values represent in ASCII?

$$62\quad 69\quad 6E\quad 61\quad 72\quad 79$$

After converting these values into letters, convert them into binary numbers to see how the computer would represent this word. Hexadecimal is a sort of "language" in between the letters that people understand, and the binary zeroes and ones that the computer understands.

Problem 131. How would you represent your own name in hex and in binary code? (Note: the ASCII code for the " " blank space character is 32 decimal.)

3-D Color Cube

All important data stored within a computer must ultimately be displayed on the screen to be meaningful to the user. This data may be displayed alphanumerically (numbers and characters) or graphically (colorful pictures). Thanks to the continuous improvements in graphical technology, colorful pictures have become the way of the future. But several issues are involved in displaying graphics on a computer.

One important element is color. How does a computer understand and represent all the different colors that it displays on the screen? Several color representation models are used by the computer industry. However, one particular model known as the RGB model has been adopted as the computer graphics standard.

Color classifications were developed in the 18th century to represent every color within the visible spectrum. One significant color group, the "primary colors" of red, green, and blue (RGB), is important because almost every color hue can be matched as a unique combination of these three colors. This RGB scheme is mathematically modeled as a three-dimensional unit cube with the RGB components mapped into x, y, and z coordinates. So a color at location (x, y, z) within this cube is comprised of the following primary color mixture:

$$\text{color} = (x, y, z) = (\text{R\_component, G\_component, B\_component})$$

The computer stores the RGB values for each colored pixel (picture element or dot) on the screen in graphics memory. Then a special-purpose graphics controller board converts these digital values into analog video signals, and displays the result on the screen. The more memory bits or "precision" available per pixel to store the RGB values, the more colors the computer system is capable of displaying. For example, graphics memory with 24 bits per pixel (eight bits for each primary component) can display up to 2^{24} (over 16 million) possible colors.

Problem 132. This is a unit cube and therefore the x, y, and z values are between zero and one. It is possible to describe all shades of red as $(x, 0, 0)$ where x varies between zero and one. An x value very close to zero, like 0.0001, would be very faint pink. An x value very close to one, like 0.9999, would be bright red, almost full-intensity red. What would you suppose are the coordinates of black (no color at all) and white (full color)?

Problems 133. What are the (x, y, z) coordinates, at full intensity, of each of the three primary colors: red, green, and blue?

Problem 134. When the primary colors were classified, it was discovered that not all hues could be directly derived from RGB. However, one of the primary colors could be added to this hue to match a combination of the remaining two primary colors. This property resulted in the "secondary colors" of cyan, magenta, and yellow (CMY), that are considered complements of the primary colors respectively. In other words, cyan is a mixture of green and blue, magenta is a mixture of red and blue, and yellow is a mixture of red and green.

What are the (x, y, z) coordinates, at full intensity, of each of the secondary colors?

Problem 135. How would you express all monochrome (grey) shades?

Polly Moore
Mathematics and Computing

I have always enjoyed math, and took my first calculus course as a senior in high school. When I went to college, I was still trying to decide whether to major in math or chemistry, so I took a number of courses in both areas. I got a lot of encouragement from my math professors, and organic chemistry seemed particularly abstruse. In the end the decision was easy: math!

At the time, I wanted to teach math, so I earned a masters degree and a PhD. Unfortunately, when I graduated there were very few university positions for mathematicians, and even postdoctural jobs were difficult to find. So, toward the end of my graduate school career I looked hard at what applied mathematicians in industry were doing, and even interviewed a dozen of them to find out what kinds of math they used most frequently. In addition to finishing my degree, I also took courses in numerical analysis, differential equations, and computer programming. Not surprisingly, all these courses were taught in the School of Engineering, not in the Math Department! (That situation has since changed, I'm happy to report.)

After I finished my PhD, I went to work in Aerodynamics Research for the Boeing Company. I had one (very hard) problem to solve, involving fluid flow around a radially symmetric object (like an engine housing). It was interesting mathematically, but I found that I didn't really have an affinity for airplanes or aerodynamics. However, my old chemistry interest was still there, so when I had an opportunity to work for Merck, a large pharmaceutical company, I took it. My role was as a mathematical consultant to the research organization. I worked on a variety of problems in chemistry and biology—modeling fermentations, developing algorithms in molecular modeling, data-modeling problems in various kinds of assays, and so on. This application of mathematics was much more enjoyable. I learned a lot about drug development and practical mathematics!

After two years at Merck, my husband (a scientist there) and I made a major career decision to join Genentech. At the time, it was a risky decision because biotechnology was not yet proven as a viable source of commercial drug products. But sometimes you need to take a chance, so we moved to California, and proceeded to work harder than we ever had before! Genentech wasn't large enough to need a fulltime mathematician, so I worked with the computing group instead. As an applied mathematician, I had used computers, but never before had been in charge of keeping them running for others. Quite an education. Over time, I moved into the management side of computing, and eventually became vice president of all the computing activities in the company. The only disadvantage to this job is that I rarely get to work on interesting math problems anymore (budgets don't count as interesting!) But I've been able to help a new company grow and become successful, so it's been worth the trade-off. I suspect that if and when I retire, I will somehow get involved with math again!

Cell Division Timing

Genetic engineering makes use of many different types of cells to produce proteins from DNA. For example, human insulin and human growth hormone are produced this way. Cells are grown in flasks or fermenters, and they grow and multiply when they get enough food and oxygen. Because cells multiply by dividing in half, one single cell gives rise to two new ones. The new cells inherit the same DNA as the parent cell, so genetic information is passed on from a few cells in the lab to many cells in the manufacturing plant.

The cells multiply at regular intervals, which are different for different kinds of cells and for the same kind of cells grown under different conditions. Knowing the time interval is an important part of planning an experiment, as well as a handy way to summarize information about a particular cell line.

This problem is how to calculate how long it takes for a cell to divide if you know something about how many cells are present at various times.

Problem 136. Suppose you know the number of cells at each of the following times:

Time in hours (t)	Number of Cells
0.0	1
0.5	2
1.0	4
1.5	8

What's the formula for the number of cells at any time t?

Problem 137. Now suppose that you start with 100 cells instead of just one. The data will look like:

Time in hours (t)	Number of Cells
0.0	100
0.5	200
1.0	400
1.5	800

What's the formula for the number of cells at any time t? Using the notation $C_0 = 100$ (the number of cells at time 0.0), can you rewrite the formula using C_0?

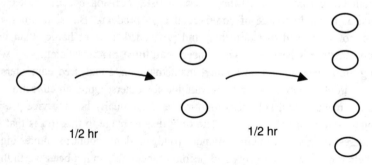

1/2 hr   1/2 hr

FIGURE 54
Cells multiplying at regular intervals

Problem 138. The general formula for the number of cells at time t is:

$$C(t) = C_0 2^{kt}$$

where t stands for time, C_0 is the number of cells at time 0.0, and $1/k$ is called the *doubling time*. What is the doubling time in the example in Problem 137?

Problem 139. Suppose you know that at time $t = 0.0$ there are 10,000 cells and three hours later, at time $t = 3$, there are 160,000 cells? What is the doubling time for these cells?

Problem 140. (Harder) Using the same example as in Problem 139, suppose that at time $t = 3$, there are only 150,000 cells. What is the doubling time now?

Note: one way to solve this problem involves using $\log_2$ of a number. Since $\log_2$ is not normally found on a calculator, but $\log_{10}$ is, it is useful to know they are always related by the equation:

$$\log_2(x) * \log_{10}(2) = \log_{10}(x)$$

Lynn Stiglich

Electrical Engineering

When I was in high school, I didn't know what I wanted to do for a career. Science—especially astronomy, nature, the human body, and archeology—was always fascinating to me, so I knew I wanted to go into a technical field. During all four years in high school, I studied math, including algebra, geometry, trigonometry, advanced math, and introductory calculus.

Once in college, I decided to go to medical school and become a doctor, so I started the pre-med program while also working parttime. The competition to get into med school was fierce, and I soon realized my grades were not good enough. One weekend, I went to the annual Engineering Fair, presented by the School of Engineering. There I saw an experiment in biomedical engineering that was fascinating. A digital-analog (or hybrid) computer was modeling the human circulatory system. Students used this computer model to simulate heart attacks, arrhythmias, and other heart problems. Later, I talked with several engineering professors, and realized I could get a degree in electrical engineering, specializing in biomedical engineering. In this way, I could work in a medically-related field. My decision was made. I went on to major in electrical engineering at the University of Wisconsin in Madison.

In college, I studied algebra, calculus, differential equations, linear algebra, and physics because they were prerequisites for the engineering courses. Computer science and statistics were also required. Almost all of the engineering courses required math to solve the problem sets, so I really did use all the math I learned! Since I still had to earn money, I got a job in a computer lab, and later as a student technician in the Plasma Physics Department because I wanted to get hands-on experience in a lab environment. I learned how to do everything from ordering electrical components to using machine tools. Even more importantly, I developed self-reliance and confidence that I could get the job done. This work experience was incredibly valuable after I got my degree and started to work as an engineer in the real world.

While earning my BS-EE degree, I learned there are many other interesting fields of electrical engineering. In addition to biomedical engineering there are power, optics, radio frequency and microwave communications, semiconductors, computers, plasma, and microprocessors, just to name a few. After applying to many companies, I interviewed with Hewlett Packard (HP) and was offered a job there. I realized it was an excellent company with lots of different career opportunities, so I accepted the job and moved to Palo Alto, California.

At first, I worked as a production engineer in the Microcircuits Department. My job was to solve technical problems that came up in the manufacturing of our products. The products were radio frequency (RF) and microwave test equipment. My favorite project was getting a microwave cavity-tunable oscillator into production. I developed the test process, computer test program, documentation, and production area.

After that, I went on to several other jobs, one as a production section manager for microwave switches. Each job required technical expertise, some knowledge of the business, and the ability to communicate effectively with people and in writing.

Life was not all work! I met a man, fell madly in love, and got married. He was also an engineer at HP. We transferred to another HP division in Santa Rosa, California. My next job was to design and build a test chamber (described below) and learn all about the tests. Next, I went to work in the Marketing Department as a product marketing engineer. In this job, I learned about new microwave technologies, defining new products, doing market research, and what products our customers wanted. This was a great job!

Then my husband and I started a family. I took a long maternity leave and stayed home for a year, to be with our baby girl. Then I came back to the Marketing Department at HP and took a job as a sales development engineer. I answered questions and solved problems for our field sales engineers, so they could sell HP products to customers who needed them.

We added to our family once again, and currently I am a mom and a "domestic engineer." I still find my math and engineering knowledge to be very useful, and plan to return to the workplace in the future. I know there will be new technologies and products to learn, and plenty of ways to apply everything I learned in college and past jobs.

In summary, there are many different fields of electrical engineering. There are also many different companies and types of jobs. I have worked for HP for my entire career, and have had many different jobs there. Depending on your interests and abilities, there are jobs in many departments: Research and Development, Marketing, Production, Field Sales and Quality. They all require a degree in Electrical Engineering or some other technical area, and they all use math in everyday work.

Testing for Radiated Interference

HP builds equipment and instruments that are used to measure and test other instruments. Many engineers and scientists use HP test equipment in their daily work to perform experiments and take measurements. There are laws and regulations that HP must meet to make sure that its instruments perform properly. One law requires that instruments must not produce unwanted energy or radiation that would interfere with the operation of other instruments. An example of this is when you turn on a hair dryer or an electric motor and it makes the television picture fuzzy.

This form of unwanted energy is called interference and it can travel through space or over electrical wires. When it travels through space it is called radiated interference and when it travels through electrical wires it is called conducted interference. HP performs tests on the equipment it builds to make sure this will not happen. The following problems are about the test for radiated interference. The first problem derives a formula, and the next two problems use the formula in numerical calculations.

A diagram of the test setup used to measure this interference is shown in Figure 55. The equipment under test (EUT) is placed on a table at a height a above the ground. An antenna is placed a distance L away from the EUT at a height h above the ground. The antenna measures the radiated interference coming from the EUT to make sure it is under the test line limit.

This test must be performed in a large metal test chamber or shielded room to make sure the antenna does not pick up interference from some other source, like another instrument or a radio station. Signals outside the chamber cannot penetrate the metal. The interference is measured as it comes in a direct path from the EUT to the antenna. Some of the energy will bounce off the ceiling of the test chamber and reach the antenna, causing an incorrect reading. So the height of test chamber H is designed to cut down on this reflected energy. In addition, energy-absorbing foam shaped in pyramidal cones is glued onto the ceiling, to cut down on the reflected energy.

If the angle of incidence A is kept less than 45 degrees, the reflected energy from the ceiling will not cause an incorrect reading. This is because for angles of A less than 45 degrees, the incident energy is absorbed directly by the cones or reflected further into the cones, where it is absorbed. With angle A more than 45 degrees, any energy reflected off the ceiling is much more difficult to absorb or attenuate, and can contribute to the direct radiation being measured with the antenna, causing an incorrect reading.

Problem 141.   Find an expression for the angle of incidence A in terms of the chamber height H, the antenna height h, the EUT height, and distance L. The engineer designing the shielded room can use this expression to optimize the room's dimensions—especially height H—to minimize unwanted reflections. The design engineer must choose between cost and performance. A shielded room with a ceiling that is higher than necessary is more expensive to build. But the ceiling must be high enough to keep the angle of incidence A less than 45 degrees.

Hint.   Draw a mirror image of the test setup diagram using the ceiling as the reflective plane. This technique is used in solving optics problems as well. Identify the similar triangles that allow angle A to be expressed in terms of H, h, a, and L.

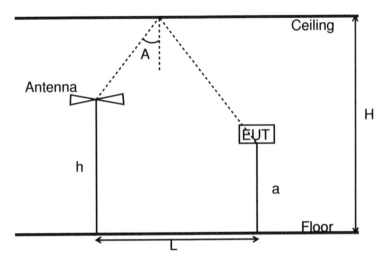

FIGURE 55
Test setup diagram

Angle A will be expressed as the arctan of an expression using H, h, a, and L.

If you get stuck on one part, peek at that part of the answer, then continue to work out the rest of the answer on your own. The answer is broken down into seven steps.

Problem 142. The Federal Communications Commission (FCC) test setup for radiated interference requires that the antenna and EUT be placed one meter apart. The heights of both the antenna and EUT must be one meter above the floor. Using the formula, show that a shielded room with a ceiling height of two meters will be adequate for performing the radiated interference test. Refer to Figure 55 for the test setup.

Problem 143. A European regulatory agency requires a test setup slightly different from the FCC test setup. For this test, the distance between the antenna and the EUT is three meters. The height of both the antenna and EUT is one meter.

Two shielded rooms are available for the test. Chamber #1 has a ceiling height of two meters, and Chamber #2 has a ceiling height of 12 feet. To determine which chamber is acceptable for testing the EUT, calculate the angle of incidence A for both rooms. Hint: one yard $= 0.9$ meters.

Nancy G. Roman

Astronomy

As long as I can remember, I wanted to be an astronomer. In fifth grade, I started an astronomy club among my friends, and proceeded to read every astronomy book I could find in our local library. In high school, I happily took both math courses—geometry and algebra. In college, I majored in astronomy, but also took five years each of math and physics. The only way I managed to get through history and German in my freshman year was because I did not need to spend much time studying math and astronomy. I received a BA in Astronomy from Swarthmore College in 1946, and a PhD in Astronomy from the University of Chicago in 1949.

In my thesis on the Ursa Major Cluster (this cluster includes all but the end stars in the Big Dipper), I studied the motions of many stars to determine which ones were most likely to be members. I used membership in the cluster to determine distances and the brightnesses the stars would have if they were at a standard distance. Then I applied this same method to other clusters, to calibrate the brightnesses of stars from the appearance of their spectra. Astronomers use this method to estimate the distance to stars that are too far away to measure their apparent change in direction as the earth orbits the sun.

From 1949 to 1955, I worked at the University of Chicago, first as a research associate and then as an assistant professor. I became interested in stars with abnormally large velocities and proved that even among the bright stars near the sun, stars with fewer heavy elements move more rapidly. I also taught graduate courses and did a great deal of observing.

I enjoyed teaching very much, but did not believe that I had a chance for tenure at the University of Chicago. In fact, no women had tenure in a major astronomy department at that time, and very few had faculty positions. Fortunately, things are very different today. There are many women in senior positions on university faculties, and the director of the National Optical Astronomy Observatories is a woman.

I was offered an interesting position in the new field of radio astronomy at the Naval Research Laboratory. From 1955–1959, I worked on various projects, including measuring the distance to the moon by radar and mapping the sky at 67 cm. I also continued my research on stellar motions. In 1956, an astronomer who had been born in Russia was invited to the dedication of a new observatory in Soviet Armenia. He declined, fearing that he might not be able to leave the country again once he was there. As a result of the observatory director's interest in a short paper I had written, I was invited to take his place.

When NASA was formed in 1958, most of its science staff came from the Naval Research Laboratory. Although I had few scientific contacts with these particular groups, I was well known as a result of my trip to the Soviet Union. They asked if I would like to set up a program in space astronomy. The challenge of starting with a clean slate, defining a program that would influence astronomy for decades, was too great to turn down, even though I wasn't sure about becoming an administrator. However, I actually enjoyed it. People are more difficult and complex than stars, but often more interesting.

At NASA, I was responsible for those investigations in astronomy, solar physics, geodesy, and relativity that required making observations and conducting experiments in space. In this role, I organized a coherent program of scientific investigations, balancing the technical capabilities available and which projects members of the scientific community wished to pursue. Then I convinced senior administrators to support the program, and worked with engineers to help them understand the scientific requirements. Although I prepared testimony for Congress, I never testified myself.

Science is an international activity, particularly in astronomy which has a small number of participants and little commercial value. I met frequently with representatives of the European Space Research Organization to coordinate our programs and avoid duplication. A number of satellites were designed by Europeans but launched by the U.S., and included U.S. experiments. Europeans also flew experiments on our satellites. For many years, I was the highest-ranking woman in NASA. The only other female scientist was a woman astronomer I hired, at first part-time when her children were young, and eventually fulltime. Now, there are many professional women in NASA.

I retired from NASA in 1980 to maintain a home for my elderly mother. Since then, I have kept active with parttime positions. I provided support for various NASA programs, particularly the Hubble Space Telescope and the Earth Observation System. Since 1980 I have worked at the NASA Space Science Data Center where I edit and document astronomical catalogs for electronic archiving in the Astronomical Data Center. These catalogs can be retrieved by astronomers throughout the world in machine-readable form. For this project, I worked closely with my counterparts in France, exchanging catalogs and techniques. Last year, I spent a week observing for the first time in 18 years, and enjoyed it as much as ever.

Throughout history, there have been a comparatively large number of women in astronomy. Several made major contributions to the field. However, until my generation, they were usually restricted to low-level positions, doing a great deal of drudgework of the sort now given to computers. Being a pioneer sometimes presented challenges but was always exciting.

Galactic Orbits

This problem is from the time I still did research. (Analyzing budgets requires math, but is not as interesting.) I had to solve this type of problem to understand how the nature of stars changed with the characteristics of their galactic orbits.

One step in the problem requires spherical trigonometry. Other than the given formulae, no math is required beyond plane trigonometry. A terrestrial globe, marked in latitude and longitude, may be helpful for visualizing the necessary geometry. I have broken the solution into steps, and suggest that each be treated as a separate subproblem. The main challenge of this problem is the logical thinking necessary to see the required steps and particularly, to keep the signs correct.

Definitions, formulae, and assumptions:

Just as latitude and longitude are used to locate places on earth, so astronomers use similar coordinate systems to locate celestial objects. For this problem, we will be studying the place and motion of a star in the Milky Way galaxy. For many problems, astronomers think in terms of an imaginary celestial sphere surrounding the earth. We use a three-dimensional polar coordinate system, called galactic longitude and latitude, denoted by l and b respectively, to locate a star on this sphere. On earth, latitude $= 0$ is the terrestrial equator, and longitude $= 0$ is the meridian in Greenwich, England. On the celestial sphere, $b = 0$ is the plane of the galaxy, and $l = 0$ is the meridian in the galactic center direction. See figure 56.

Astronomers measure brightness in a very old system of units called magnitudes. These units are logarithmic. That means, if there are two stars and the brighter one is 100 times as luminous as the fainter one, they will differ in magnitude by five magnitudes, with the fainter one having the larger magnitude.

That is, if two stars have the magnitudes m_2 and m_1, their luminosities (brightnesses) , l_1 and l_2 are such that :

$$m_2 - m_1 = 2.5 \log(l_1/l_2)$$

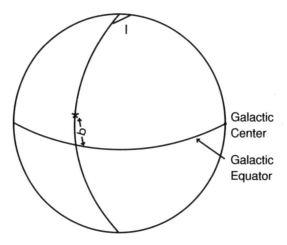

FIGURE 56
The galactic coordinate system

For example, if two stars have $m_2 = 5$ and $m_1 = 2.5$, $\log(l_1/l_2) = 1$. Hence, star m_2 is ten times brighter than star m_1.

Just as the planets move in orbits around the sun, stars move in orbits around the center of the galaxy (the center of mass of all stars and other material within its orbit). The mass of a planet can usually be neglected in respect to that of the sun, and the mass of a star can be neglected in respect to the mass of the galaxy. The orbit is a conic section, almost always an ellipse. The ellipse is defined by its semimajor axis (half of the long axis), and its eccentricity (the difference in the lengths of the long and short axes divided by the length of the long axis). The semimajor axis and the eccentricity are both determined by the velocity of the star and its distance from the galactic center. The smallest distance from the galactic center is called the perigalactic distance, the largest distance is called the apogalactic distance. Obviously, the perigalactic distance is $a * (1 - e)$ and the apogalactic distance is $a * (1 + e)$. See figure 57.

The sun moves in a circular orbit, with a speed of 235 km/sec around the center of the galaxy, at a distance of $3 * 10^{17}$ km. The velocity is directed toward $l = 90°$, $b = 0°$.

The velocity of a star can be resolved into two components. The first is a velocity along the line of sight to the sun, called "radial velocity". The second is a velocity on the plane of the sky, perpendicular to the line of sight, called "tangential velocity". The actual measurement of the motion perpendicular to the line of sight is of the apparent motion in seconds of arc, and is called "proper motion". If the distance of the star is known, the proper motion can be

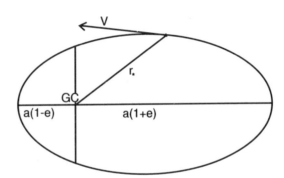

FIGURE 57
Orbit of a star around the galactic center

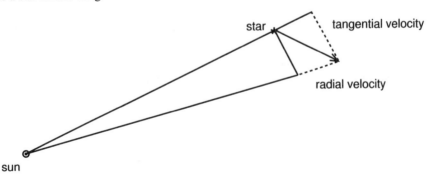

FIGURE 58
Radial and tangential motion

converted into tangential motion measured in kilometers/second. The radial velocity is positive if it is directed away from the sun. The direction on the celestial sphere it points toward must be given for the tangential velocity.

For most problems, we can describe the orbit of a star by the projection of the orbit on the galactic plane and its inclination. Using scaler values, the orbit is determined by two equations:

$$V_\parallel^2 = GM(2/r_* - 1/a)$$

and

$$(r_* * V_\perp)^2 = GM * a * (1 - e^2)$$

Notation:

l = galactic longitude
b = galactic latitude
V = the total velocity of the star
r_* = the distance of the star from the galactic center
a = the semi-major axis of the star's orbit
M = the mass of the galaxy
G = the gravitational constant
e = the eccentricity of the orbit
$V_\perp$ = the velocity of the star perpendicular to the direction to the galactic center
$V_\parallel$ = the velocity of the star parallel to the direction to the galactic center
m = magnitude
l = luminosity
i = inclination of the orbit
h = height of the star above the galactic plane

The section that uses spherical trigonometry (section D) requires the following equations:

$$\cos a = \cos b * \cos c + \sin b * \sin c * \cos a \quad \text{(law of cosines)}$$

$$(\sin \alpha)/(\sin a) = (\sin \beta)/(\sin b) = (\sin \gamma)/(\sin c) \quad \text{(law of sines)}$$

where a is the side opposite angle α, b is the side opposite angle β, and c is the side opposite angle γ.

Problem 144. A star at $l = 125$, $b = 10$ has an apparent magnitude of 8.5. If the star were at $3 * 10^{14}$km, it would have a magnitude of -1.5. The star has a radial velocity of -100km/sec, and a proper motion of $0''.02$/year toward $l = 300$, $b = -20$. What are the semimajor axis, eccentricity, and inclination of the star's galactic orbit? What are the star's perigalactic and apogalactic distances? Use only velocities in the plane to describe the orbit except for the inclination.

The problem can be broken down into smaller steps:

A. Determine the distance to the star from the sun. Use the magnitude at $3 * 10^{14}$ km and the actual magnitude to determine the actual distance.
B. Determine the position of the star with respect to (a) the sun and (b) the galactic center.

C. Determine the tangential velocity of the star.

D. Resolve the tangential velocity into components perpendicular to the galactic plane and in the plane. Resolve the latter into components perpendicular and parallel to the galactic center. Drawing the geometry on a grapefruit may help with this section.

E. Resolve the tangential velocity into components perpendicular to the galactic plane and, in the plane, perpendicular and parallel to the direction to the galactic center.

F. Add the components from the radial and tangential velocities to compute the three components of the star's total motion.

G. Compute the total velocity in the plane of the galaxy.

H. Determine the characteristics of the star's orbit. Since the gravitation constant and the mass of the galaxy are the same for the star and sun, the characteristics of the sun's orbit can be used to determine the constants in the equations.

Note. None of the observed quantities is very accurate. Therefore, three significant figures normally suffice.

Claudia Zaslavsky

Author

By the time I was four years old I was using mathematics in a practical way, counting out and selling three or four candies for a penny in my dad's store. A few years later I operated the cash register in our clothing store. My parents' trust in me and praise from our customers, as well as my own interest, encouraged me to see math as my field. Everyone expected me to major in mathematics, and I did.

I have worked in many fields, mostly related to mathematics. In college I majored in statistics, and then earned a master's degree in actuarial mathematics, the kind of mathematics that is used in the insurance field. However, since insurance companies were not hiring women at that time, I worked as an accountant and as a junior engineer until I had children. Afterwards I went into teaching—high school and teacher education—until my retirement. I continue to conduct courses and workshops for teachers.

I taught mathematics in a school district in New York State that was known nationwide for integrating its schools by busing. In the sixties students could take courses like African history and Swahili, and a college professor conducted an after-school course in African history for teachers in the district. To satisfy the course requirement for a report on some relevant subject, I decided to write about African mathematics, but was unable to find books on the subject in any library. Therefore I had to write the book myself! To do this, I had to read a great deal in the fields of anthropology, history, and many other subjects. The title is *Africa Counts: Number and Pattern in African Culture,* and it is now available in paperback.

I became interested in the mathematics that people in various societies have developed to satisfy their own needs and interests. This mathematics may be very different from school math, but it gave rise to school math. Students, and even teachers, do not realize how much mathematics was developed by the peoples of Africa, Asia, and the Americas. Just imagine the level of mathematics the Egyptians had to develop almost five thousand years ago in order to build their pyramids. Two thousand years later, Greek scientists were completing their higher education in Egypt and Asia. Our place-value system of written numerals originated in India and was transmitted to other parts of Asia, northern Africa, and the Mediterranean region by the Arabs. It was not until several centuries later that this system was widely adopted in Europe. The Maya of Central America and Mexico may have been the first to use a symbol for zero and had a very advanced calendar.

Later I wrote several books for children, a book for parents called *Preparing Young Children for Math,* and an activities book for middle grades, *Multicultural Mathematics: Interdisciplinary Cooperative-Learning Activities.* My book for adults, *Fear of Math,* was published by Rutgers University Press in June, 1994, and I am writing two books for teachers on multicultural mathematics.

Rules for Three-in-a-Row Games

I will describe a problem that I had to solve when I was writing my book *Tic Tac Toe and Other Three-in-a-Row Games, from Ancient Egypt to the Modern Computer.*

Most young people have played Tic-Tac-Toe. Two players take turns placing an X or an O in any of the nine spaces of the three-by-three game diagram. The goal is to be the first to make a straight line of three X's or three O's. After a while the game may become boring because the person who starts cannot lose and the person who goes second cannot win, unless one of the players is careless.

Actually a great deal of mathematics can be learned from this simple game. For example, think about the two opening moves. The first player can place an X in any of nine different spaces. For each move by Player One, Player Two has a choice of eight spaces in which to place an 0. It might seem that there are $9 * 8$, or 72 ways to make the two opening moves.

In fact, there are only 12 ways to make the two opening moves. To prove that statement, make a model. On a blank sheet of paper draw the game diagram with a dark marker, so that you can see it on the other side (but be careful not to mark your desk or table). Then place an X in a corner and an O next to it (but not in the center). By turning the game board a quarter-turn each time, you have four different opening moves. Then flip the paper to the other side and repeat the procedure—four more opening moves. All eight positions are really the same.

Analyzing the moves by flipping and turning the game board in this way depends upon a mathematical idea called symmetry. Can you figure out in how may different ways the two opening moves can be made? The answer is twelve. Try to find them.

Three-in-a-row gameboards were found on the roofing tiles of an ancient Egyptian temple constructed about 3,300 years ago. Since then they have spread all over the world, in many versions—to other parts of Africa, to Asia, to Europe, and to the Americas. My goal was to collect some of the many different ways of playing three-in-a-row games and to write a book for young people in which I described these ways. A publisher agreed to publish the book. An artist drew suitable illustrations for each game.

One of the books I used in my research was *Games of the North American Indians,* by Stewart Culin, originally published in 1907. Culin, an anthropologist, was particularly interested in the games of non-European people. His book had a whole section on three-in-a-row games among different Native American groups. Among these groups were the Pueblo people of New Mexico. The word "pueblo" is Spanish, and means both "people" and "town." On page 798 Culin wrote: "A boy from Isleta, named J. Crecencio Lucero, described the people of this pueblo as playing a board game which they called picaria (Spanish, pedreria), little stone. They used diagrams of two kinds." However, he was not very specific about the rules of the game.

Some years later I had a wonderful experience in connection with this game. At a conference I met a math teacher from Isleta Pueblo, who invited me to visit his school. Before my visit I sent him a copy of the pages in Culin's book describing the game. When I arrived, bringing a copy of my games book, he introduced me to his sixth-grade class. Then, pointing to one student, he said, "It was probably your grandfather who described a game in this book." This boy had the same last name as the boy who had given the information to Stewart Culin almost a century earlier! It seems more likely that he was the student's great-grandfather.

Several recent game books include a three-in-a-row game called Picaria. According to the rules given in these books, the game is played on thirteen points (I will explain that later). I knew about similar games, called mill or three-men's morris, that were played on nine points, in Spain, England, and other European countries, but never one played on thirteen points on this type of board.

This is what I wrote in my book *Tic Tac Toe and Other Three-in-a-Row Games from Ancient Egypt to the Modern Computer,* published in 1982 by T. Y. Crowell. (Reprinted with the author's permission.)

PICARIA

COUNTERS: One player has three white counters, the other player has three black counters.

GAME BOARD: Copy this square diagram. The game is played on the nine marked points.

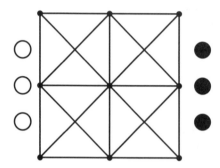

FIGURE 59
The 9-point Picaria Diagram

HOW TO PLAY: The two players take turns placing one counter at a time on an empty point on the board. When all six counters have been placed, the players take turns moving one counter at a time along any line to the next empty point. Jumping over a counter is not allowed.

OBJECT: Each player tries to make a row of three counters of the same color. A row can be made across, up and down, or diagonally. Altogether there are eight ways to win. They are the same as in Tic-Tac-Toe.

FINISH: The winner is the first player to make a row. If neither player can get three in a row and the game becomes boring, call it a draw.

Changing the Rules.   Some people play Picaria on the 13 points marked on this game board. Follow the rules above, but with these differences:

1. Neither player may place a counter in the center of the board until all six counters are on the board.

2. You can make three in a row anywhere along a diagonal, as long as no empty point is included. There are 16 different ways to make a row. Besides the three horizontal rows and three vertical rows, there are three possible ways along each of the two diagonals and four ways along the lines connecting adjacent sides of the large square.

Problem 145.   Draw the 12 ways to make two opening moves in Tic-Tac-Toe.

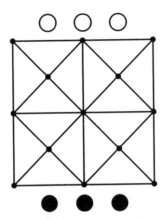

FIGURE 60
The 13-point Picaria Diagram

Problem 146. Play Picaria on the nine-point game board.

Problem 147. Play Picaria on the thirteen-point game board.

Problem 148. Compare the two versions of Picaria. Discuss with your classmates which is the better game. You might consider such factors as the time it takes to play the game, the degree of challenge (you don't want a game that gets boring very quickly), and whether the person who goes first is more likely to win than the one who goes second. You can probably think of other factors after you have played a few games using each set of rules.

Problem 149. Once you have mastered the basic game of Picaria, you can try using four counters for each player on either the nine-point or the thirteen-point game board.

Jean E. Taylor
Mathematics

I didn't know I liked math until I took the Kuder Personal Preference Test and a course in algebra. A typical question in the Kuder Test is: would you rather grow a flower, sell a flower, or develop a new breed of flower? The test told me that my interest in computation was at the 96th percentile. Discovery of my interest in math (according to the test) surprised me because I had thought that math was boring and also because no one else in my family was interested in math or science. Then I discovered algebra. It gave me a powerful tool to solve the kinds of puzzles I enjoyed doing anyway, so I was convinced that math could really be interesting and useful.

I took algebra in the 9th grade, geometry in the 10th, algebra II in the 11th, and two math courses in the first semester of my senior year—trigonometry and solid geometry—and calculus the second semester (a rigorous, epsilon-and-delta course!). I always got A's in math. The only sciences I took were chemistry and physics. I was nervous about taking physics, because I thought it was a "boy's subject," but my counselor convinced me to try it and I became the best student in the class. I also took two semesters of drafting in high school, one of them in summer school; that remains one of my proudest accomplishments, since I went from F's on all my early work up through D's, C's, B's, and finally A's.

I went to Mount Holyoke College because I wanted to go East to college, having grown up in California, and at that time it seemed to me that the best colleges in the East that accepted women were women's colleges. I did not specifically want to go to a women's college, but it turned out well for me because I learned there to "take myself seriously" and to plan on a research career.

When I went to college, I was debating between mathematics and chemistry as a major (and planned to become a high school teacher). But during the first year, I found math boring again—I was required to repeat my high school calculus. Chemistry, on the other hand, was challenging—in fact, I got advanced placement in chemistry purely on the basis of my math skills! After toying with the idea of majoring in psychology, I finally decided to major in chemistry, although I did take two years of calculus and a year of abstract algebra so I'd know about symmetry groups and the like. After a fascinating course in Parties and Politics, I almost spent a summer as an intern for Senator Alan Cranston in Washington, DC, but at the last moment opted for an NSF (National Science Foundation) program in chemistry research instead.

After two years of graduate school in chemistry, during which I had passed all the required PhD exams but only begun my thesis work, I switched to mathematics. I found the subject of differential geometry (curves and surfaces) absolutely fascinating! Since I had to work very

156

hard to catch up, it helped that I had taken or audited several math courses (including complex analysis, algebraic topology, and differential geometry) while I was a chemistry grad student. Throughout my graduate study I had an NSF graduate fellowship, and I am really grateful to NSF because that fellowship stayed with me through my change of subject and then two changes of university.

I eventually got my PhD in mathematics at Princeton University. I was an instructor at MIT for a year; I've been at Rutgers University ever since then, and I've been a full professor since 1982.

My old interest in chemistry and physics now manifests itself in the type of mathematical problems I work on, shapes of surfaces of crystals, and I use a computer to do my "drafting" as part of The Geometry Center. Articles about my work have appeared in *Scientific American* (July 1976, October 1993, November 1993), *Science News,* and other magazines and books; I have also written and published nearly 60 papers and two videotapes.

I have two step-children (I joined the family when they were about 10) and another daughter. All three of my children are interested in mathematics. The older two already have PhDs and work in applied mathematics, whereas the youngest began taking math classes at Princeton University when she was in the ninth grade and is at this time still in high school. Maybe our kids like math because my husband (also a mathematician) and I talk about it all the time and communicate our enthusiasm to our kids.

I enjoy lots of things other than mathematics, like skiing, scuba diving, wind-surfing, kayaking, hiking, and mountain climbing. I used to be a pretty good rock-climber, and I even learned how to fly an airplane. But it is in mathematics research that I enjoy the intellectual freedom of trying to make new connections, develop new theories, and understand previously unexplained phenomena.

Thinking Geometrically

This problem is one of the foundations of geometry, and is related to such fundamental ideas as curvature. It can also lead one into symmetry groups and thus algebra, and so forth. That's one of the reasons I really like it. It was this type of question—why things are the shapes that they are—that lured me into mathematics and continues to fascinate me.

This is an open-ended problem designed to get the reader involved in thinking geometrically and away from thinking that mathematics is mostly formulas.

Problem 150. Make sheets of paper that are covered by equilateral triangles of the same size (you could copy Figure 61) and then cut them out . What happens when you tape or glue them together around a point? What is different about the behavior when you tape 3 or 4 or 5 together at a point, compared to taping 6 together, and what is different again when you tape 7 or more together?

Problem 151. How about if you use squares instead of triangles? Regular pentagons (5 sides)? Regular hexagons (6 sides)? "Regular" means all sides are equal, like a square. Regular pentagons are shown in Figure 62. Note that 6 of the equilateral triangles in Figure 61 form a regular hexagon.

Can you determine what the condition is for the taped-together shapes to lie flat? To form something that is very floppy? To form a fairly rigid tent that cannot be flattened?

Problem 152. Now see what shaped bodies you get when 3 equilateral triangles meet at every vertex, or 4 equilateral triangles, or 5 equilateral triangles. Can you form a single body without any free edges when 6 or more triangles meet at each vertex? What happens when you use all squares instead of triangles? Or all regular pentagons?

Problem 153. Make "hyperbolic paper" by taping together 7 equilateral triangles at every vertex. Note how floppy it is! Also note how many triangles you have to use—pick a vertex as a "center" and color the 7 triangles around it red, then color all the triangles that touch those 7 triangles blue, and then all the triangles that touch a blue triangle yellow. How many blue triangles are there, and how many yellow triangles? How fast is the number of triangles growing? How does it compare to what happens with flat paper where you have 6 triangles around every vertex?

Problem 154. I gave Problem 153 above to the students in my graduate course in differential geometry. One of the students got so tired of cutting out and taping together triangles that he decided to find out the minimum number of cuts you have to make in regular paper in order to construct hyperbolic paper. He came up with a very ingenious answer. You might want to try to work on that problem too. He is a very bright grad student, but he did not have to use any advanced mathematics to formulate or solve his problem, just a lot of thinking and experimenting.

FIGURE 61
Equilateral Triangles

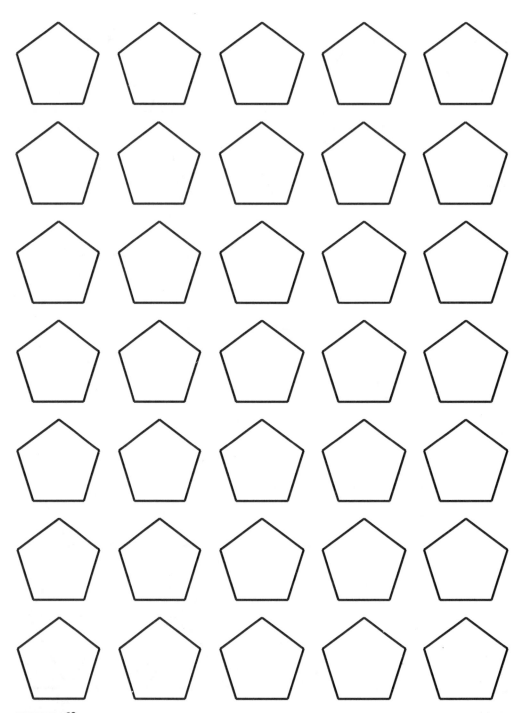

FIGURE 62
Regular Pentagons

Reflections on WAM

Eileen L. Poiani
Founding and National Director of WAM, 1975–1981

In 1975, the Mathematical Association of America (MAA) asked me to develop a lecture-ship program that would encourage young women to study mathematics.The problem, a dispro-portionately small number of women entering the mathematical sciences and other mathematics-dependent careers, had been identified in the early 1970's. A contributing factor was the fact that American young women tend to avoid taking high school mathematics after tenth grade, but young men normally study four years of high school mathematics.

At the time I was asked to direct the program, I was also coordinating the Speakers Bureau of the Saint Peter's College Mathematics Department. This bureau sent department members to make presentations on mathematical topics in New York and New Jersey metropolitan-area secondary schools to stimulate student interest in mathematics. Through the Speakers Bureau, I stressed that mathematics is the "critical filter" in career access, as sociologist Dr. Lucy Sells pointed out. To be prepared for college-level calculus and qualify for majors that lead to careers in the sciences, medicine, engineering, and other technical fields, young women should study at least four years of high school mathematics .

Saint Peter's is a Jesuit college founded in 1872. As the first woman to join the Mathematics Department, I was sensitive to the special needs of women students, who were admitted to the College beginning in 1967. During the 1970's, women accounted for close to half of the College's mathematics majors.

With this background, I was especially pleased and honored to have been asked to launch what would become the *Women and Mathematics: WAM* program. Funded in 1975–76 by a $7,500 grant from IBM, the program was planned to be national in scope, beginning with a region on each coast. Sponsored by the MAA, WAM was under the auspices of the Committee on Secondary School Lecturers, chaired by Professor Donald B. Small, always a strong supporter of WAM's goals.

For the 1975–76 academic year, I served as both National Director and Coordinator of the New York/New Jersey region. Professor Jean P. Pedersen of Santa Clara University de-veloped an active and successful West Coast WAM region in the San Francisco Bay Area. A Greater Chicago regional coordinator was also appointed in 1975, but illness postponed the establishment of that region for another year.

Programs like WAM, with a mammoth goal and a minimum budget, must rely on the ded-ication, time, and energy of its participants. Long evenings and weekends were spent planning school visits and answering the voluminous correspondence that were generated by every an-nouncement about WAM. Our initial publicity, for instance, prompted inquiries from 36 states, as well as Washington, D.C., Puerto Rico, and Canada. In its first year, the program reached more than 6,000 students plus 500 teachers and guidance counselors. Those figures more than doubled the following year, when 115 schools were visited. By the 1980–81 academic year, WAM had ten active regions with 160 speakers, and by conservative estimates had reached over 72,000 students, 875 schools, and 9400 teachers, counselors, parents, and other adults. The rest is history.

Early in the life of the program, WAM decided that, to provide role models, all its speakers should be women. While academic women are encouraged to be speakers, we also seek speakers in other walks of life to talk about the importance of mathematics as preparation for job and career entry and advancement. A typical WAM visit consists of an interesting mathematical talk, an informal dialogue session with the audience, and a conversation with guidance counselors and teachers. Because every school and region of the country have unique needs and characteristics, the success of each visit relies heavily on personal tailoring.

Initially, WAM programs focused on the tenth grade, and gradually moved into the lower grades. This was because studies had shown that young women drop out of mathematics after just two years. Also, speakers felt more comfortable speaking to a high school, rather than to a middle- or elementary-school, audience.

Identifying effective speakers for WAM is a critical and time-consuming process. Dr. Henry O. Pollak, President of the MAA from 1975–76, has been a friend of WAM since its inception. He helped me meet with women at Bell Laboratories to recruit speakers. As a result, Susan M. Devlin, Mary-Jane Cross, and Lorraine Denby of Bell Laboratories were my successors as coordinators of the New York/New Jersey region. Mrs. Devlin subsequently served as Special Materials Coordinator, developing a useful speakers package, and Ms. Denby served as New Projects coordinator.

In the early years, MAA President Henry L. Alder, Secretary David P. Roselle, and Executive Director Alfred B. Willcox and his staff, headed by Elaine Pedreira, gave invaluable support and encouragement to the WAM team. Our liaisons to IBM—Dr. Thomas R. Horton, Corporate Director of University Relations, and Miss Frances M. Kelly, Manager of Special Educational Support Programs—were loyal and generous friends to WAM. I also owe special gratitude to Prof. John Ernest from the University of California at Santa Barbara for donating a sizable supply of reprints of his article, *Mathematics and Sex,* published in the *American Mathematical Monthly,* October 1976, for distribution at WAM presentations.

Our goal is to gradually expand WAM, by identifying dynamic and responsible coordinators, and by retaining the personal touch needed for successful school visits. From 1975 to 1981, I convened annual sessions with the regional coordinators at the same time as the Joint Mathematics Meetings in January. These marathon sessions were notorious for their intensity, but they strengthened WAM through healthy discussion and sharing of the coordinators' experiences.

Evaluation of WAM's success was always a prime concern of mine. We did this by developing evaluation forms for students, the host organizations, and speakers. At first, the responses were tallied by hand, thanks to my mother, Eileen L. Poiani, Sr. Eventually Susan Devlin and Lorraine Denby took over this task on computer. Our 1977 paper, *The WAM Program: A Preliminary Statistical Evaluation,* looked at the short-term effect of WAM presentations by analyzing pre- and post-student attitudes towards mathematics and sex roles in math-related fields at seven urban and suburban schools. This pilot study led to a doctoral dissertation by Dr. Carole B. Lacampagne, who followed me as National Director of the program.

Knowing that guidance counselors are pivotal in influencing students, the New York/New Jersey WAM region designed a special event on December 5, 1978, called *MODE: Math Opens Doors Everywhere*. Funded by the Prudential Foundation and held at its Newark, N.J. headquarters, this successful day-long event attracted 100 counselors. The next year, it was reprised in Chicago by WAM Coordinator Dr. Kathleen Sullivan, and I hope it will be repeated in the future.

WAM's limited financial resources require continuous thrift. These resources were multiplied manyfold by the speakers, who often waived their modest honoraria, and by contributed services. The program owes much to the talented women who gave so generously because they believed in its mission. WAM has indeed been a cooperative effort.

The lectureship program served as the model for the Blacks and Mathematics (BAM) program, which was founded in 1977 with a grant from the Exxon Foundation. BAM sought to encourage African-American middle-school students to continue with mathematics and consider careers in math-related areas. The need to attract more minority students to the mathematical sciences remains a priority of the MAA.

Even before the term "networking" was popular, WAM recognized the importance of working with other intervention programs. For example, WAM speakers participated in the *Expanding Your Horizons* programs; in the *Futures Unlimited* conferences; and in Connecticut's *Project to Improve Mastery of Mathematics* (PIMM).

From the original one-page reference list I prepared in 1976, to the extensive bibliography developed with input from all WAM regions, the program has distributed thousands of bibliographic lists over the years. Today's multifaceted reference is *Sex Differences and Anxiety Relative to Mastering Mathematics: An Annotated Bibliography*.

WAM has played a special role in my life, leaving me many memories along with copies of correspondence and materials that fill a file cabinet in my home. Every week, inquiries from students at all levels of education continue to arrive, asking for information on mathematics anxiety, careers in mathematics, references for research papers and dissertations, and courses of study.

When I accepted the directorship, I said that intervention programs like WAM will know they have succeeded when they put themselves out of business. That time may be drawing a little nearer, but it has not yet arrived. Work remains to be done because, as Plato said in *The Dialogues:* "Nothing can be more absurd than the practice which prevails in our country, of men and women not following the same pursuits with all their strength and with one mind, for thus the state, instead of being a whole, is reduced to a half."

Rena Haldiman, astronaut crew training instructor

Susan Loverso, software engineer at Thinking Machines Corporation, with her son Peter

Donna McConnaha Sheehy, civil engineer, with her husband, Ed, and their children, Min and C. J.

Jean Taylor, Professor of Mathematics at Rutgers University

Lynn Stiglich, electrical and "domestic" engineer, with her daughter Larissa

Jill Baylor, electrical engineer

Sally Lipsey, retired professor of mathematics and current book author, with her grandaughter Anna

Rosalie Dinkey, retired chemical engineer

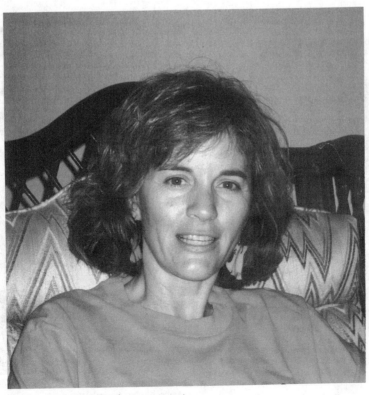

Marilyn K. Halpin, oil-and-gas accountant

Fahmida N. Chowdhury, professor of electrical engineering at Michigan Technological University

Eileen Poiani, Professor of Mathematics and Assistant to the President for Planning at St. Peter's College

Shelley J. Smith, archaeologist for the Bureau of Land Management

Helen Townsend-Beteet, pharmacist, and her husband Michael Beteet

Nancy Powers Siler, dietitian

Elaine Anselm, training coordinator at Xerox,
with her daughter Ruth

Beth MacConnell, fish pathologist for the US Fish and Wildlife Service

Amy C. R. Gerson, electrical engineer for Boeing

Caroline P. Nguyen with colleagues from the Hubble Space Telescope Program

Martha Leva, professor of business courses at the Ogontz Campus of Penn State University, with her daughters Dana and Nicole

Susan Knasko in front of the Monell Chemical Senses Center where she does research on taste and smell

Claudia Zaslavsky, author

Mary Campione, software engineer, with her
husband Richard in Monterey, California

Polly Moore, Vice President at Genentech

Maryam Hastings, Professor of Mathematics and
Computer Science at Marymount College in
Tarrytown, New York

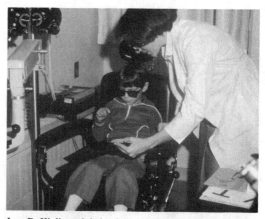

Jane D. Kivlin, ophthalmologist, with a patient

Renate McLaughlin, Professor of Mathematics at the University of Michigan-Flint

Nancy G. Roman, astronomer

Linda Valdés, Assistant Professor of Mathematics at San Jose State University, with her husband (on her right) and a friend

Sharon Lum, electrical engineer for IBM

Barbara Swetman, computer scientist, with her daughter Amanda

Eileen Thatcher, a professor at Sonoma State University, with her son Adrian

Solutions

Knasko—Environmental Psychology

Problem 1. 5.2 (Add the scores of all the expert subjects exposed to an unpleasant odor and divide by the total number of subjects in that group: $8 + 9 + 0 + 3 + 6 = 26$, $26/5 = 5.2$)

Problem 2. Beginners exposed to an unpleasant odor (total number of errors $= 113$, for an average of 22.6 errors).

Problem 3. No. The average number of errors for the experts ranged from 3.2 for those exposed to a pleasant odor, to 5.2 for those exposed to an unpleasant odor. The average number of errors for the beginners ranged from 14.4 for those exposed to no odor, to 22.6 for those exposed to an unpleasant odor.

Problem 4. Odor seems to have more effect on beginners. The average number of errors made by experts exposed to odors differed only by 1.2 (more errors) or .8 (fewer errors) compared to experts not exposed to an odor. On the other hand, exposure to an odor led to more errors, on average, for beginners than exposure to no odor (14.4 errors on average) in the no-odor condition. That is 8.2 fewer errors than beginners exposed to an unpleasant odor, and 7.8 fewer errors than beginners exposed to a pleasant odor.

Problem 5. Under all odor conditions, experts (as a group) made fewer errors on this paper-and-pencil task than beginners. Neither pleasant nor unpleasant odors seem to influence the performance of experts. However, both types of odors seem to cause beginners to make more errors.

Problem 6. No. The average mood score for both groups was four.

Problem 7. Yes. It seems that for both experts and beginners, the trend in mood was the same. Subjects exposed to a pleasant odor tended to be in a better mood than subjects exposed to either unpleasant or no odors. There was also very little difference in the mood of subjects exposed to unpleasant or no odors.

Problem 8. No. Not unless a *very small* difference in the number of errors was important.

Problem 9. Experts exposed to a pleasant odor. Experts in the three conditions don't differ in performance, but those exposed to a pleasant odor are in a better mood. Beginners exposed to a pleasant odor are also in a better mood, but they make more errors than beginners exposed to no odor.

Problem 10. Many responses are correct!

Campione—Software Engineering; Computer Science

Problem 11. There are many ways to store kerning values so only important data is stored, and all other values are assumed to be zero.

One way is to store the data in a $3 * nn$ matrix (where nn is the number of kerning pairs) as follows:

	1	2	3	...	nn
1st character	$a1$	$b1$			$c1$
2nd character	$a2$	$b2$	...		$c2$
Kerning value	$v1$	$v2$			vnn

Or you could save things in a single-dimension array like this:

	$a1$	$a2$	$v1$		$b1$	$b2$	$v2$		...

Without knowing the exact contents of the table, we can't specify how much memory will be saved or required. But we can analyze it. For instance, assume the initial matrix contained 90 percent zeros and ten percent useful information. At first glance, you might decide that both of the solutions above saved 90 percent of the original storage space. However, they used three cells per kerning pair—not one. (The indices into the arrays no longer supply us with any pertinent information.) So, we really require 30 percent of the initial space. That means neither of the solutions is better than the other, except when it comes to their algorithms for retrieving stored information.

Problem 12. The palindrome problem lends itself to a solution utilizing a "stack." (In computer science, a commonly-used data structure is called a stack.) A stack is like a stack of books—when you place books on the top of it, the stack grows taller. When removing books, you take them from the top of the stack. This is commonly referred to as "last in first out."

So in the case of the palindrome, place the letters of the phrase or sentence onto a stack, one by one, counting them as you go. That's how to determine the number of letters in the palindrome. At this point, begin removing letters from the stack, until you reach the halfway point. Then, print out either the single center character (if the palindrome has an odd number of letters) or two center characters (if the palindrome has an even number of letters).

Again, there are no "right" answers in computer science. If you came up with another solution and it works—then it is just as correct as mine.

The answer to the second part of the problem is: if the palindrome has an even number of characters, then the two characters located in the center must be identical. Otherwise, the palindrome would be different when read backward, so it wouldn't be a palindrome at all!

Problem 13. I've used psuedo-code to illustrate the answer. Remember there is never just one answer in computer science. If yours is different from the one shown here and it works, then it's also correct.

```
/*
 * Using recursion compute n factorial
 */

subroutine rfactorial (input integer "n")
returns integer
{
        if n is equal to 1
            return 1
        else
            return (n * rfactorial (n-1) )
}
/*
 * Using iteration to compute n factorial
 */

subroutine ifactorial (input integer "n")
returns integer
{
        initialize result = 1
        initialize i = n
        while i is greater than 1 {
            result = result * i
            i = i - 1
        }
        return result
}
```

Problem 14. I've used psuedo-code to illustrate the answer. Remember, there is never just one answer in computer science. So if yours is different from the one shown here and it works, then it's also correct.

```
/*
 * Using recursion compute the Fibonacci numbers
 */

subroutine fibonacci (input integer "n")
returns integer
    {
```

```
        if n is less than or equal to 2
        then return 1;
        else return fibonacci(n-1) + fibonacci(n-2)
}
```

It is difficult to illustrate the use of a subroutine on paper. To calculate the seventh Fibonacci number, pretend you are the computer executing the subroutine. You are "called" with $n = 7$, like this:

```
call fibonacci(7)
```

and you must return an integer answer. Since this is a recursive subroutine, you'll have to call yourself several times to calculate the answer. The first time you go through the subroutine, you'll return `fibonacci(6)` + `fibonacci(5)`, but how do you know what they are? To find out, you must call `fibonacci(6)`, and then `fibonacci(5)`, and so on.

The only time you'll ever be able to return an integer without recursively calling the routine is when you are called with n less than or equal to two. Those calls allow you to escape from the recursion and start to "bubble up" the answers, until you get back to the answer for `fibonacci(7)`.

In programming a recursion, one of the most common bugs occurs when, for some reason, the escape condition is never true. With this subroutine that is not possible, because it's so simple. But if the escape condition is very complicated, it's easy to make a mistake and create an escape condition that will never be true for some input data. Then, there is no way to escape the recursion. The subroutine will literally call itself forever, never returning any value at all. When this happens on a computer, you have to figure out an external way to make the program stop executing. On some simple systems, the only way to do that is—turn off the power!

Smith—Archaeology

Problem 15. From Figure 1, the sides of the dig to be mapped are six and eight meters, so the length of the third tape measure, from the $(0,0)$ origin to the north-east corner stake, can be calculated using the Pythagorean theorem:

$$Z^2 = X^2 + Y^2$$

$$Z^2 = 6^2 + 8^2$$

$$Z = R(36 + 84) = R(100) = 10 \text{ meters}$$

Problem 16. From Figures 2 and 4, the middle ordinate, M, is 2.1 cm. The chord length, C, is 10 cm. Putting these numbers into the formula to calculate the radius gives:

$$R = C^2/8M$$

$$R = (10)^2/8(2.1)$$

$$R = 100/16.8 = 5.95 \text{ cm.}$$

So the circumference is $2\pi R = 2(3.14)(5.95) = 37.37$ cm.

Problem 17.

Sample A

Plant Pollen	Counts	Total in Sample	% of Total
Pine	300	6383	30
Sagebrush	226	4806	22.6
Grass	275	5851	27.5
Oak	72	1532	7.2
Spruce	100	2128	10
Cactus	27	575	2.7
Total	1000	21,275	100

Sample B

Plant Pollen	Counts	Total in Sample	% of Total
Pine	290	4,028	29
Sagebrush	240	3333	24
Grass	300	4167	30
Oak	60	833	6
Spruce	90	1250	9
Cactus	20	278	2
Total	1000	13,889	100

The "total in sample" is a proportion problem, calculated to solve for x, where n is the pollen count of a particular plant.

$$\frac{470}{10,000} = \frac{n}{x}$$

For pine, as an example:

$$\frac{470}{10,000} = \frac{300}{x}$$

$$470x = 300 * 10,000$$

$$x = \frac{3,000,000}{470} = 6383.$$

The "% of total" is calculated by dividing the particular plant count by the total count and multiplying by 100:

$$\frac{n}{N} = \frac{300}{1,000} * 100 = 30\%$$

The percentages of each plant pollen grains are not significantly different between Samples A and B.

The total pollen in Sample B is 65% of the total in Sample A.

$$\frac{13,889}{21,275} * 100 = 65\%$$

Hastings—Mathematics and Computer Science

Problem 18.   Let X be a point obviously outside the curve. If you start at point A in Figure 7 and travel to X, you cross the curve an even number of times. If you travel from B to X, you cross the curve an odd number of times. Therefore, A is outside the curve, and B is inside the curve. Apply the same procedure to determine all other points. For Figure 7, the answers are (from left to right):

A outside, B inside, C inside

A outside, B inside, C outside

A outside, B inside, C outside

Problem 19.   Curve 8 is a four-connected, simple closed curve, since every point on the curve is four-adjacent to exactly two other points on the curve. Curve 9 is an eight-connected, simple closed curve, since every point on the curve is eight-adjacent to exactly two other points on the curve. Curve 10 is neither. It is not a four-connected, simple closed curve, because point A has only one four-adjacent point on the curve. It is not an eight-connected, simple closed curve, because point B has three eight-adjacent points on the curve—points $B1$, $B2$, and $B3$. Can you find another point on curve 10 that has three eight-adjacent points on the curve? There are five more such points.

Stout further proved that a four-connected, simple closed curve with at least eight points divides the computer screen into two eight-connected regions called components. An eight-connected, simple closed curve with at least four points divides the computer screen into two four-connected components.

Problem 20.   Ms. Vos Savant answers, "Yes." She explains her answer this way, "The first door has a one-third chance of winning, but the second door has a two-thirds chance of winning."

With the first choice, the contestant has a two-thirds chance of picking the wrong door. If she agrees to switch after one of the losing doors is revealed, then either of the two (original) losing choices must lead to a win (upon the second choice). So the probability of losing on the first choice (two-thirds) becomes the probability of winning on the second choice—if the contestant agrees to switch!

In general, when choosing one of the three doors, the probability of picking the wrong door is two-thirds. By switching after one of the doors is revealed, two-thirds of the time you will change from a loss to a win situation. Ms. Vos Savant suggests that readers set up experiments if they are not satisfied with her explanation.

This problem and the response it created are an interesting illustration of how counter-intuitive mathematical problems can be. Some of the reaction to this famous problem and solution may also be due to a misunderstanding of the explanation. Although Ms. Vos Savant is recognized as having an exceptionally high IQ, some readers appeared to be unconvinced largely because of her perceived lack of mathematical "authority." Among the interesting comments received were, "Maybe women look at math problems differently than men," and, "You're wrong, but look at the positive side; if all those PhD's were wrong, the country would be in very serious trouble." The last comment was made by a PhD from the Army Research Institute. No wonder we have been having such serious troubles!

Sheehy—Civil Engineering

Problem 21.   Alternative 1. Route 1-4-5-6-3

Segment	Miles		$Cost/Mile		$Cost
1-4	0.2	*	$ 5,000	=	$1,000
4-5	0.5	*	$15,000	=	$7,500
5-6	0.3	*	$10,000	=	$3,000
6-3	0.3	*	$12,000	=	$3,600
Surfacing	1.3	*	$10,000	=	$13,000
Cattleguards	3 each	*	$ 3,200	=	$ 9,600
Right of way	2 cases	*	$20,000	=	$40,000
			Total Cost	=	$77,700

Alternative 2. Route 1-2-3

Construction Cost:

Segment	Miles		$Cost/Mile		$Cost
1-2	1.3	*	$25,000	=	$32,500
2-3	1.0	*	$17,500	=	$17,500

Surfacing Cost:

Segment	Miles		$Cost/Mile		$Cost
1-2	1.3	*	$15,000	=	$19,500
2-3	1.0	*	$10,000	=	$10,000

Drainage Cost:

Segment	Crossings		$Cost/Crossing		$Cost
1-2	1	*	$7,000	=	$7,000

Total Cost = $86,500

Alternative 1 is recommended for access to Section 21. The route is the least expensive (by $8,800) even with right-of-way costs, and has the least environmental impact. It avoids a major crossing and the additional surfacing requirements.

Note: This example only has six road segments, but a typical network analysis problem may contain several hundred of them.

Problem 22. As presented, this is a simple cost trade-off problem with no present-worth calculations; therefore, you must compare the total costs of doing the job. Two different methods for solving this problem are presented: A and B. They yield the same answer.

Method A. Eight-hour day:

$$\text{Crew days} = \frac{640 \text{ person-hours}}{4 \text{hrs/day} * 4 \text{ persons}} = 40 \text{ crew days}$$

$$\text{Cost} = 40 \text{ days} * \$18.00/\text{hr.} * 8 \text{ hrs/day} = \$5760$$

Ten-hour day: For the 10-hour day, the pay is $1.5 * \$18$ for the extra two hours beyond the normal eight-hour day.

$$\text{Crew days} = \frac{640 \text{ person-hours}}{6 \text{ hrs/day} * 4 \text{ persons}} = 26.6 \text{ crew days (use 27 days)}$$

$$\text{Cost} = (27 \text{ days} * \$18.00/\text{hr} * 8 \text{ hrs/day}) + (27 \text{ days} * \$18/\text{hr} * 1.5 * 2 \text{ hrs/day})$$

$$= \$3,888 + \$1,458 = \$5,346$$

Method B. Eight-hour day:

$$\text{Daily crew cost} = 8 \text{ hours} * \$18/\text{hour} = \$144/\text{crew-day}$$

$$\text{Daily crew work hours} = 4$$

$$\text{Cost per productive hour} = \frac{\$144/\text{crew-day}}{4 \text{ hours/day}} = \$36/\text{crew-hour}$$

$$\text{Total cost} = 640 \text{ person-hours} * \frac{\$36/\text{crew-hour}}{4\text{-person/crew}} = \$5,760$$

Ten-hour day: For the ten-hour day, the pay is $1.5 * \$18$ for the extra two hours beyond the normal eight-hour day.

$$\text{Daily crew cost} = (8 * 18) + (2 * 18 * 1.5) = 144 + 54 = \$198/\text{crew-day}$$

$$\text{Daily crew work hours} = 6$$

$$\text{Cost per productive hour} = \frac{198}{6} = \$33/\text{crew hour}$$

Since 640 person-hours are not evenly divisible by four persons working six hours/day, you must round up to 648 person-hours, or 27 crew-days:

$$27 \text{ crew days} = 27 \text{ days} * 6 \text{ hrs/day} * 4 \text{ persons} = 648 \text{ person-hours.}$$

Using 648 person-hours:

$$\text{Total cost} = 648 \text{ person-hours} * \frac{\$33/\text{crew-hr}}{4 \text{ person/crew}} = \$5,346.$$

Note: on overtime, if it were possible to work 22 days at 10 hours per day and four days at 11 hours per day, the cost would be $5,256.

Problem 23. There are at least a couple of ways to approach this problem.

First, look at the cost per productive hour (on-the-ground time), by calculating the ratio of "mobilization" cost to actual work hours. This method is probably preferred, because it brings productive time into the analysis.

Second, look strictly at mobilization cost—the cost to support a crew in the field. This method may be appropriate if productive time is on the same order of magnitude.

Method A. Cost Per Productive Hour Basis

Alternative 1. Use existing permanent bunkhouse.

$$\text{Transport cost} = 80 \text{ mi/day} * \$0.25/\text{mi} * 88 \text{ days} = \$1760$$
$$\text{Travel time cost} = 2 \text{ hrs/day} * \$36/\text{hr} * 88 \text{ days} = \$6336$$
$$\text{Total Cost} = \$8096$$

$$\text{Work hours} = 88 * 8 = 704 \text{ hours}$$
$$\text{Cost per hour} = \$8096/704 = \$11.50/\text{hr}$$

Alternative 2. Use portable bunkhouses.

$$\text{Bunkhouse cost, present value} = \left[4000 * 2 - 300 * 2\left(\frac{1}{1.08^8}\right)\right]$$
$$= \left[8000 - 600(.5403)\right]$$
$$= (8000 - 324.18)$$
$$= \$7675.82$$

The yearly payment, made at the end of each of 8 years, is x. Then

$$\$7675.82 = 1.08^{-1}x + 1.08^{-2}x + \cdots + 1.08^{-8}x$$
$$= 5.7468x$$
$$x = \$1335.67 \text{ per year.}$$
$$\text{Extra maintenance cost} = \$200(2) = \$400 \text{ per year.}$$
$$\text{Moving cost} = 3 \text{ moves } (8 \text{ hrs/move})(\$36/\text{hr})$$
$$= \$864 \text{ per year.}$$
$$\text{Total cost} = \$2599.67 \text{ per year.}$$

$$\text{Work hours} = (88 \text{ days} - 3 \text{ days})(8 \text{ hrs/day}) = 680 \text{ hours.}$$
$$\text{Cost per hour} = \$2599.67/680 = \$3.82/\text{hr.}$$

Method B.   Total Cost Including Lost Time

Alternative 1.   Use existing permanent bunkhouse.

$$\text{Transport cost} = \$1760 \text{ per year}$$
$$\text{Cost of time lost} = 2 \text{ hrs/day} * \$36/\text{hr} * 88 \text{ days} = \$6336$$
$$\text{Total cost} = \$8096$$

Alternative 2.   Use portable bunkhouses.

$$\text{Annual bunkhouse cost} = \$1335.67$$
$$\text{Extra maintenance} = \$400$$
$$\text{Moving cost} = \$864$$
$$\text{Total cost} = \$2599.67$$

Obviously the portable bunkhouse is the answer for this problem.

Problem 24.
$$\text{Break-even volume} = \frac{R_H - R_L}{H_L - H_H}$$
$$R_H = \$94,000 \qquad R_L = \$55,000 \qquad H_L = \$60/\text{mbf} \qquad H_H = \$35/\text{mbf}$$
$$V = \frac{\$94,000 - \$55,000}{\$60/\text{mbf} - \$35/\text{mbf}} = 1560\text{mbf} = 1.56\text{mmbf}$$

With a planned sale volume of two mmbf, the ridgetop road is the recommended alternative.

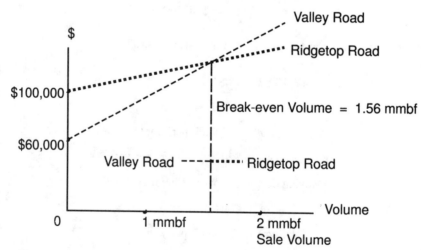

FIGURE 63
Valley Road vs. Ridgetop Road

Problem 25.   Since the clinometer measures degrees at eye-height, six feet in the air, the large triangle made by the tree and the ground must be drawn, instead of the imaginary line that is

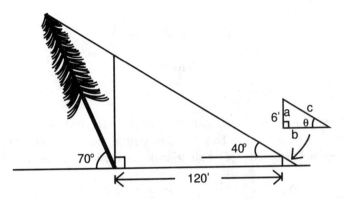

FIGURE 64
Add the Small Triangle

six feet off the ground. To do this, figure out the dimensions of the small triangle shown in Figure 64.

In Figure 64, you can see that the angle θ is 40°, just like the angle measured by the clinometer at 6′.

$$\theta = 40° \quad \tan\theta = \frac{a}{b} \quad \tan 40° = \frac{6'}{b} \quad b = \frac{6'}{\tan 40°} \quad b = 7.15'$$

Adding the 7.15′ to the 120′ gives you the dimensions of the final large triangle made by the tree and the ground, shown in Figure 65.

By subtracting 70° from 90°, you can calculate γ:

$$\gamma = 90° - 70° = 20°$$

$$B = \gamma + 90° = 20° + 90° = 110°$$

$$A + B + C = 180°$$

$$40° + 110° + C = 180° \qquad C = 130°$$

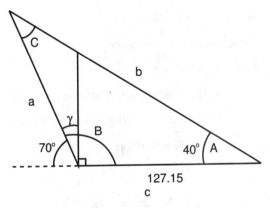

FIGURE 65
The Final Large Triangle

$$\frac{a}{\sin A} = \frac{c}{\sin C} = \frac{b}{\sin B}$$

$$\frac{a}{\sin 40°} = \frac{127.15}{\sin 30°}; \qquad \frac{a}{0.6428} = 254.3; \qquad a = 163.5'$$

So the length of the tree is 163.5'.

Problem 26. Given: Tension in lines is represented by force vectors acting in the direction of the lines. These single forces can be reduced to vertical and horizontal components. These components correspond to the sides of a right triangle such that $F_h = T \cos \theta$ and $F_v = T \sin \theta$.

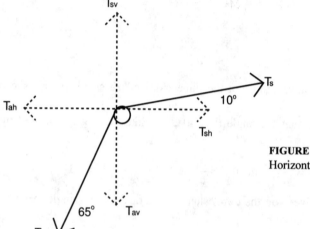

FIGURE 66
Horizontal and vertical forces of skyline

T_s = tension of the skyline

T_{sv} = vertical tension of the skyline

T_{sh} = horizontal tension of the skyline

T_a = tension of the anchor

T_{av} = vertical tension of the anchor

T_{ah} = horizontal tension of the anchor

$T_s = T_a$ (The tension in any line running over a block

is the same on both sides of the block.)

A. First determine the horizontal and vertical forces of the skyline.

$$T_{sh} = T_s \cos \theta$$

$$T_{sh} = 55,000 \text{ lbs } (\cos 10°)$$

$$T_{sh} = 54,164.43 \text{ lbs } \rightarrow$$

$$T_{sv} = T_s \sin \theta$$

$$T_{sv} = 55,000 \text{ lbs } (\sin 10°)$$

$$T_{sv} = 9550.65 \text{ lbs } \uparrow$$

This is the total vertical lift of the skyline—the amount of logs that can be hauled in one load.

$$T_{ah} = T_a \cos \phi$$

$$T_{ah} = 55,000 \text{ lbs } (\cos 65°)$$

$$T_{ah} = 23,244.00 \text{ lbs } \leftarrow$$

$$T_{av} = T_a \sin \phi$$

$$T_{av} = 55,000 \text{ lbs } (\sin 65°)$$

$$T_{av} = 49,846.93 \text{ lbs } \downarrow$$

$$\sum F_h = \text{sum of the horizontal forces}$$

$$\sum F_v = \text{sum of the vertical forces}$$

Since the forces are in opposite directions, they should have opposite signs in order to add them together. Let the force to the right $\rightarrow$ be positive and the force to the left $\leftarrow$ negative; let the force up $\uparrow$ be positive, and the force down $\downarrow$ negative.

$$\sum F_h = T_{ah} + T_{sh}$$

$$\sum F_h = (-23,244.00) + (54,164.43)$$

$$\sum F_h = 30,920.43 \text{ lbs } \rightarrow$$

$$\sum F_v = T_{av} + T_{sv}$$

$$\sum F_v = (-49,846.93) + (9550.65)$$

$$\sum F_v = -40,296.29 \text{ lbs } \downarrow$$

B.   The horizontal tension of the guyline, $T_g h$, must equal the sum of the horizontal forces of the skyline, $\sum F_h$, for the system to balance.

$$T_g = \text{tension of the guyline}$$

$$T_{gv} = \text{vertical tension of the guyline}$$

$$T_{gh} = \text{horizontal tension of the guyline}$$

$$T_{gh} = \sum F_h = 30,920.43 \text{ lbs } \leftarrow$$

$$T_{gh} = T_g (\cos \gamma)$$

$$30,920.43 \text{ lbs } = T_g (\cos 50°)$$

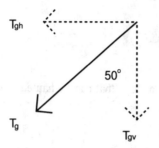

FIGURE 67
Horizontal and vertical forces of guyline

$$T_g = \frac{30,920.43}{\cos 50°}$$
$$T_g = 48,103.65$$
$$T_{gv} = T_g(\sin \gamma)$$
$$T_{gv} = 48,103.65(\sin 50°)$$
$$T_{gv} = 36,849.54 \text{ lbs } \downarrow$$

Total vertical load acting on the tree $= T_{gv} \downarrow + \sum F_v$
$$= 36,849.54 + 40,296.29$$
$$= 77,145.81 \text{ lbs } \downarrow$$

Valdés—Mathematics

Problem 27. Hardware designers must consider both efficiency and cost in their designs. They want all vertices to be reachable from each other. Spanning trees would do this, but an interruption in the flow of electricity in any one of the wires would disrupt the entire network. If each vertex were on a cycle, that would solve the problem. On the other hand, the designer would like to keep the cost down by using as few wires as possible. These cubic graphs are examples of a good design that meets both goals: efficiency and cost.

Two different graphs can be drawn:

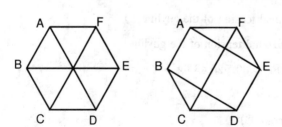

FIGURE 68
Cities connected by three phone lines

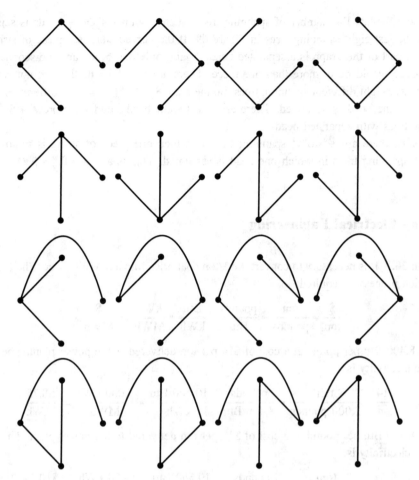

FIGURE 69
The spanning trees of the cubic graph

Problem 28. See Figure 69.

Problem 29. A spanning tree cannot contain a cycle. Cycles can be found in two different places in the family of graphs under consideration: the cycles formed with all the pieces of string, and the cycles formed within each bead.

How many beads are there and how many pieces of string? Since there are four vertices in each bead, and as many strings as beads, there are $\frac{p}{4}$ beads and pieces of string.

First consider the necklace with one piece of string removed. No more than one piece can be deleted, since if two or more pieces were removed, the necklace would be in more than one piece so it would not be a spanning tree. Look at each bead and find all the spanning trees belonging to that bead. These trees are shown as the first eight trees in Figure 69. Since there are eight ways to find spanning trees for each bead and there are $\frac{p}{4}$ beads, there are $8^{p/4}$ ways of finding spanning trees in a necklace with one string removed. So there are $\frac{p}{4} * 8^{p/4}$ different spanning trees with any one of the pieces of string removed.

Now, calculate the number of spanning trees when any one of the $\frac{p}{4}$ beads is separated. Look at the last eight spanning trees in Figure 69. If the curved edge is a piece of string, the remaining part of the graph is a separated bead. Again, only one bead can be disconnected or the necklace would be in more than one piece. There are $\frac{p}{4} - 1$ beads that are not separated and that have eight different spanning trees. So there are $8 * 8^{(p/4-1)} = 8^{p/4}$ different spanning trees when one bead is separated. There are $\frac{p}{4}$ different beads, and therefore $\frac{p}{4} * 8^{p/4}$ such spanning trees with separated beads.

Finally, there are $\frac{p}{4} * 8^{p/4}$ spanning trees in which one piece of string is removed and $\frac{p}{4} * 8^{p/4}$ spanning trees in which one bead is separated. This is a total of $\frac{p}{2} * 8^{p/4}$ spanning trees.

Baylor—Electrical Engineering

Problem 30. It is necessary to convert the $/ton coal price into a $/MWh price. The following conversion process is required:

$$\frac{\$}{ton} * \frac{ton}{pounds} * \frac{pounds}{Btu} * \frac{Btu}{kWh} * \frac{kWh}{MWh} = \frac{\$}{MWh}.$$

Coal 1: 8,400 Btu per pound at a cost of $14 per ton delivered to the power plant. The cost to generate electricity is:

$$\frac{\$14}{ton} * \frac{ton}{2,000\ pounds} * \frac{pounds}{8,400\ Btu} * \frac{10,500\ Btu}{kWh} * \frac{1,000\ kWh}{MWh} = \frac{\$8.75}{MWh}$$

Coal 2: 8,800 Btu per pound at a cost of $17 per ton delivered to the power plant. The cost to generate electricity is:

$$\frac{\$17}{ton} * \frac{ton}{2,000\ pounds} * \frac{pounds}{8,800\ Btu} * \frac{10,500\ Btu}{kWh} * \frac{1,000\ kWh}{MWh} = \frac{\$10.14}{MWh}$$

Thus, Coal 1 will be selected as it costs $8.75/MWh versus a cost of $10.14/MWh for Coal 2.

Problem 31. Total capacity $= 100$ MW $+ 50$ MW $+ 200$ MW $+ 25$ MW $= 375$ MW (the sum of units A, B, C, and D).

The peak demand in year plus $1 = 1.02 *$ peak demand in year.

Year	Peak Demand	Required Capacity	Installed Capacity	Surplus (Deficit)
this year	300.0	345.0	375	30.0
Year + 1	306.0	351.9	375	23.1
Year + 2	312.1	358.9	375	16.1
Year + 3	318.4	366.2	375	8.8
Year + 4	324.7	373.4	375	1.6
Year + 5	331.2	380.9	375	−5.9

New capacity is required five years from this year.

Problem 32. The percentage represented by each type of generating unit must be calculated before creating the graph.

Type of Unit	Number	Percent	# of Degrees Out of 360
Coal/Gas	2	1.2	4.3
Coal	34	19.8	71.3
Gas/Oil	90	52.3	188.3
Oil	26	15.1	54.3
Gas Turbine	20	11.6	41.8

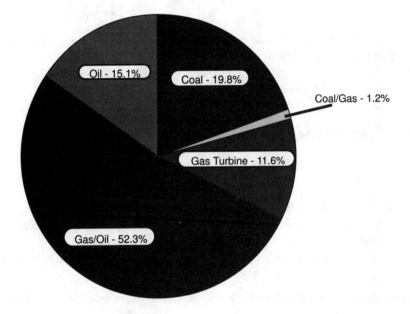

FIGURE 70
Fuels Used by Utilities Surveyed

Problem 33. The unit produced power 75 percent of 8,760 hours at 300 MW.

300 MW * .75 * 8,760 hours = 1,971,000 MWh.

Problem 34. The unit can generate 200 MW * 8,760 hours = 1,752,000 MWh in a year. It generated 1,103,760 MWh.

The ratio of these numbers is:

$$\frac{1,103,760}{1,752,000} = .63$$

The unit's capacity factor is 63 percent.

Problem 35. Part A. Capacity charges: In each month, Utility B will pay $5/KW for 100 MW. Thus, over a year Utility B will pay

$$12 \text{ months} * \$5/\text{kW} * 100 \text{ MW} * 1000 \text{ kW/MW} = \$6,000,000.$$

Energy charges: For each kilowatthour, Utility B will pay $.02. Utility B purchases 569,400 Mwh. The cost is:

$$569,400 \text{ MWh} * \$0.02/\text{kWh} * 1000 \text{ kWh/MWh} = \$11,388,000.$$

$$
\begin{array}{lll}
\text{TOTAL:} & \text{Capacity Charges} & = \$\ 6,000,000 \\
& \text{Energy Charges} & = \underline{\ \ 11,388,000} \\
& & \ \ \ \ 17,388,000
\end{array}
$$

Part B. Divide the total revenue required ($145,920,000) by the amount of energy used by the customers (1,824,000 MWh) expressed in terms of kilowatthours.

$$\frac{\$145,920,000}{1,824,000 \text{ MWh} * \dfrac{1000 \text{ kWh}}{\text{MWh}}} = \$0.08/\text{kWh}$$

Part C. Miles times the cost per mile will yield the total cost of this project.

$$46 \text{ miles} * 750,000 \ \$/\text{mile} = \$34,500,000$$

Problem 36. 1. 100 miles by truck = 100 miles $*$ 15¢/ton-mile $*$ 1$/100¢= $15 per ton
   2. 900 miles by rail = 900 miles $*$ 2¢/ton-mile $*$ 1$/100¢= $18 per ton
   3. 1000 miles where 250 miles by barge and 750 miles by rail
$$= 250 \text{ miles} * 1¢/\text{ton-mile} * 1\$/100¢ + 750 \text{ miles} * 2¢/\text{ton-mile} * 1\$/100¢$$
$$= \$17.50 \text{ per ton}$$

For this example, moving coal by truck 100 miles is cheaper than the other alternatives.

Problem 37.

$$\frac{.01 \text{ lb S}}{\text{lb coal}} * \frac{2 \text{ lb SO}_2}{\text{lb S}} * 1,000,000 \text{ Btu/MMBtu} * (1/12,000 \text{ Btu/lb})$$
$$= 1.67 \text{ lb SO}_2/\text{MMBtu}$$

This coal will generate 1.67 lb SO_2/MMBtu. The requirement is 1.2 Lbs SO_2/MMBtu.

$$(1.67 - 1.2)/1.67 = 0.28$$

Thus, 28% of the SO_2 will have to be scrubbed out of the coal.

Cominsky—Physics; X-ray Astronomy Research

Problem 38. Part 1. 6 hours (See Figure 18.)
   Part 2. 1 hour
   Part 3. P for the Earth is one year.
   Part 4. $d = 7.7 * 10^{-3}$ A.U. $= (1/1460)^{2/3}$ since six hours $= 1/1460$ year

McLaughlin—Mathematics

Problem 39. Fill in the loading table with the information provided:

	Weight (lbs)	Moment (lb.-ins./1000)
1. Basic empty weight (includes unusable fuel and full oil)	1661	63.8
2. Usable fuel (at 6 lbs/gal)[maximum: 62 gal]		
3. Pilot and front passenger	395	
4. Rear passengers	340	
5. Baggage area 1 [maximum: 200 lbs]	60	
6. Baggage area 2 [maximum: 50 lbs]	15	
7. Ramp weight and moment	2471	sum of numbers above
8. Fuel allowance for engine start and taxi	−8	−0.4
9. Take-off weight and moment (add lines 7 and 8) [maximum: 2650 lbs]	2463	

Part a. Without fuel, the plane weighs 2,463 pounds. Therefore, 187 pounds of fuel can be added before it reaches the maximum take-off weight of 2650 pounds. At six pounds per gallon, this amounts to just over 31 gallons.

Part b. Fill in the 187 pounds for fuel and look up all the moments in Figure 24 (p. 39). Weight is shown on the vertical axis and moments/1000 on the horizontal axis. There are lines for the various categories. The table above now looks like this:

	Weight (lbs)	Moment (lb.-ins./1000)
1. Basic empty weight (includes unusable fuel and full oil)	1661	63.8
2. Usable fuel (at 6 lbs/gal)[maximum: 62 gal]	187	8
3. Pilot and front passenger	395	14.5
4. Rear passengers	340	25
5. Baggage area 1 [maximum: 200 lbs]	60	5
6. Baggage area 2 [maximum: 50 lbs]	15	2
7. Ramp weight and moment	2471	117.9
8. Fuel allowance for engine start and taxi	−8	−0.4
9. Take-off weight and moment (add lines 7 and 8) [maximum: 2650 lbs]	2650	17.5

On the diagram of the envelope (Figure 25, p. 40), look up the point with x-coordinate 117.5 and y-coordinate 2650. Since the point lies on the top boundary of the envelope, this is an acceptable configuration. The center of gravity is within safe limits.

Part c. The plane has 31 gallons of fuel. Two gallons are needed to start the engines, taxi to the runway, and perform all the mandated pre-flight checks.

Since climbing to altitude requires a lot more fuel than cruising in level flight, set aside eight gallons to reach 3000 feet (the amount of fuel needed for climbing is specified in the operating handbook of each airplane). This leaves 21 gallons. With a fuel burn of nine gallons per hour, the plane can fly for 2 1/3 hours (two hours and 20 minutes). Air speed is 130 knots, so the plane can travel 303 1/3 nautical miles in two hours and 20 minutes. This translates into about 349 statute miles. But be cautious: The air speed is the same as the ground speed only if there is no wind. Suppose there are headwinds or tailwinds? Forecasts of wind speeds and directions are notoriously unreliable, even in this day of weather satellites. How far can the plane go with 31 gallons of fuel before the tanks must be filled again? (Read the comments at the end of this set of answers.)

Part d. As if the uncertainty of headwinds was not enough complication, consider another restriction: To keep our pilot's license, we must obey the Federal Aviation Administration rules about landing with enough fuel to fly for at least another 30 minutes at normal power settings. In this case that is about five gallons. So instead of 21 gallons, there are only 16 gallons remaining. With 16 gallons, the plane can fly for 16/9 hours (about one hour, 45 minutes). If the winds are calm, it could travel $130 * 16/9 \approx 231$ nautical miles, or about 266 statute miles. All the warnings about possible surprises in the form of unforeseen headwinds still apply.

This problem points out that the skill required to fly an airplane is only one small aspect of the overall picture. The pilot's judgment can have a greater effect on the safety of the passengers. There is a saying that there are old pilots and bold pilots, but no old, bold pilots! Most pilots have personal rules that they will not break; for example, some will not fly with less than one hours' worth of fuel in the tanks, some prefer 1/4 tank as reserve. In this case, I would want to land with at least one hours' worth of fuel (nine gallons) in the tanks. This leaves 12 gallons for cruising flight, about one hour and 20 minutes. Assuming no wind, the plane could travel about 173 nautical miles, or about 199 statute miles. Since headwinds are a fact of life, I would not plan a trip of more than 180 statute miles.

The moral of this problem is, that to get any utility from an airplane you need large fuel tanks. Also, try to travel light so you can load the airplane with fuel, not luggage.

Problem 40. Fill in the loading table with the information provided. Initially, put the passenger in the front seat.

	Weight (lbs)	Moment (lb.-ins./1000)
1. Basic empty weight (includes unusable fuel and full oil)	1661	63.8
2. Usable fuel (at 6 lbs/gal)[maximum: 62 gal]	372	
3. Pilot and front passenger	360	
4. Rear passengers		
5. Baggage area 1 [maximum: 200 lbs]	150	
6. Baggage area 2 [maximum: 50 lbs]		
7. Ramp weight and moment	2543	sum of numbers above
8. Fuel allowance for engine start and taxi	−8	−0.4
9. Take-off weight and moment (add lines 7 and 8) [maximum: 2650 lbs]	2535	

a) The total take-off weight would be 2,535 pounds, 115 pounds below the limit. As far as weight is concerned, you can take the passenger and all the luggage.

b) Check whether the center of gravity falls inside the envelope. Fill in all the moments with the passenger in the front seat:

	Weight (lbs)	Moment (lb.-ins./1000)
1. Basic empty weight (includes unusable fuel and full oil)	1661	63.8
2. Usable fuel (at 6 lbs/gal)[maximum: 62 gal]	372	18
3. Pilot and front passenger	360	13.5
4. Rear passengers		
5. Baggage area 1 [maximum: 200 lbs]	150	14
6. Baggage area 2 [maximum: 50 lbs]		
7. Ramp weight and moment	2543	109.3
8. Fuel allowance for engine start and taxi	−8	−0.4
9. Take-off weight and moment (add lines 7 and 8) [maximum: 2650 lbs]	2535	108.9

Since the point $(108.9, 2535)$ lies inside the envelope (Figure 25, p. 40), the passenger could sit in the front seat.

Now, see whether your passenger could sit in the back seat. Fill in the loading table again:

	Weight (lbs)	Moment (lb.-ins./1000)
1. Basic empty weight (includes unusable fuel and full oil)	1661	63.8
2. Usable fuel (at 6 lbs/gal)[maximum: 62 gal]	372	18
3. Pilot and front passenger	130	5
4. Rear passengers	230	17
5. Baggage area 1 [maximum: 200 lbs]	150	14
6. Baggage area 2 [maximum: 50 lbs]		
7. Ramp weight and moment	2543	117.8
8. Fuel allowance for engine start and taxi	−8	−0.4
9. Take-off weight and moment (add lines 7 and 8) [maximum: 2650 lbs]	2535	117.4

The point $(117.4, 2535)$ lies in the envelope. So, your passenger has a choice of front or rear seat.

Problem 41. Fill in the loading table with the information provided. Initially, put the heavier passenger in front with the pilot.

	Weight (lbs)	Moment (lb.-ins./1000)
1. Basic empty weight (includes unusable fuel and full oil)	1661	63.8
2. Usable fuel (at 6 lbs/gal)[maximum: 62 gal]	372	
3. Pilot and front passenger	320	
4. Rear passengers	50	
5. Baggage area 1 [maximum: 200 lbs]	200	
6. Baggage area 2 [maximum: 50 lbs]	50	
7. Ramp weight and moment	2653	sum of numbers above
8. Fuel allowance for engine start and taxi	−8	−0.4
9. Take-off weight and moment (add lines 7 and 8) [maximum: 2650 lbs]	2645	

Part a. As far as weight is concerned, this flight will be legal.

Part b. Now fill in the moments with the heavier passenger in front:

	Weight (lbs)	Moment (lb.-ins./1000)
1. Basic empty weight (includes unusable fuel and full oil)	1661	63.8
2. Usable fuel (at 6 lbs/gal)[maximum: 62 gal]	372	18
3. Pilot and front passenger	320	12
4. Rear passengers	50	3.5
5. Baggage area 1 [maximum: 200 lbs]	200	19
6. Baggage area 2 [maximum: 50 lbs]	50	6
7. Ramp weight and moment	2653	122.3
8. Fuel allowance for engine start and taxi	−8	−0.4
9. Take-off weight and moment (add lines 7 and 8) [maximum: 2650 lbs]	2645	121.9

The point $(121.9, 2645)$ lies in the performance envelope, so the heavier passenger can sit in front.

Part c. Now fill in the weight and balance table, with the child in front and the heavier passenger in the rear.

	Weight (lbs)	Moment (lb.-ins./1000)
1. Basic empty weight (includes unusable fuel and full oil)	1661	63.8
2. Usable fuel (at 6 lbs/gal)[maximum: 62 gal]	372	18
3. Pilot and front passenger	200	7.5
4. Rear passengers	170	12.5
5. Baggage area 1 [maximum: 200 lbs]	200	19
6. Baggage area 2 [maximum: 50 lbs]	50	6
7. Ramp weight and moment	2653	126.8
8. Fuel allowance for engine start and taxi	−8	−0.4
9. Take-off weight and moment (add lines 7 and 8) [maximum: 2650 lbs]	2645	126.4

The point $(126.4, 2645)$ lies outside the performance envelope. If the plane took off in this configuration, you would not be able to control it and would probably crash. So, the heavier passenger must sit in front.

Problem 42. Begin by filling in the weight and balance table. Fill the two baggage areas to their limits, and put the remaining supplies in the back seat.

	Weight (lbs)	Moment (lb.-ins./1000)
1. Basic empty weight (includes unusable fuel and full oil)	1661	63.8
2. Usable fuel (at 6 lbs/gal)[maximum: 62 gal]	372	18
3. Pilot and front passenger	170	6
4. Rear passengers	197	14.5
5. Baggage area 1 [maximum: 200 lbs]	200	19
6. Baggage area 2 [maximum: 50 lbs]	50	6
7. Ramp weight and moment	2650	127.3
8. Fuel allowance for engine start and taxi	−8	−0.4
9. Take-off weight and moment (add lines 7 and 8) [maximum: 2650 lbs]	2642	126.9

As far as the total weight is concerned, the flight can proceed. But the point $(126.9, 2642)$ lies outside the envelope, meaning the center of gravity is too far back. There are many different ways to redistribute the supplies: move items from Baggage Area 2 to the back or front seats; move items from Baggage Area 1 to the back or front seats; or move items from the back to the front seat. Which option would involve the least change? (Time is valuable, and you want to spend as little energy and effort on reloading the plane as possible.)

The biggest change in the center of gravity occurs if weight is moved from Baggage Area 2 to the front seat. Test the effect of moving 20 pounds from Baggage Area 2 to the front seat:

	Weight (lbs)	Moment (lb.-ins./1000)
1. Basic empty weight (includes unusable fuel and full oil)	1661	63.8
2. Usable fuel (at 6 lbs/gal)[maximum: 62 gal]	372	18
3. Pilot and front passenger	190	7
4. Rear passengers	197	14.5
5. Baggage area 1 [maximum: 200 lbs]	200	19
6. Baggage area 2 [maximum: 50 lbs]	30	3.5
7. Ramp weight and moment	2650	125.8
8. Fuel allowance for engine start and taxi	−8	−0.4
9. Take-off weight and moment (add lines 7 and 8) [maximum: 2650 lbs]	2642	125.4

The point $(125.4, 2642)$ is still outside the envelope, but the center of gravity has moved forward. The pilot probably needs to spread out charts on the front seat, so no more items should be moved there. The next best approach is to move the remaining items from Baggage Area 2 into the back seat. Here is the effect:

	Weight (lbs)	Moment (lb.-ins./1000)
1. Basic empty weight (includes unusable fuel and full oil)	1661	63.8
2. Usable fuel (at 6 lbs/gal)[maximum: 62 gal]	372	18
3. Pilot and front passenger	190	7
4. Rear passengers	227	17
5. Baggage area 1 [maximum: 200 lbs]	200	19
6. Baggage area 2 [maximum: 50 lbs]		
7. Ramp weight and moment	2650	124.8
8. Fuel allowance for engine start and taxi	−8	−0.4
9. Take-off weight and moment (add lines 7 and 8) [maximum: 2650 lbs]	2642	124.4

The point $(124.4, 2642)$ still lies outside the envelope, so move more baggage forward. Try moving 50 pounds from Baggage Area 1 to the back seat:

	Weight (lbs)	Moment (lb.-ins./1000)
1. Basic empty weight (includes unusable fuel and full oil)	1661	63.8
2. Usable fuel (at 6 lbs/gal)[maximum: 62 gal]	372	18
3. Pilot and front passenger	190	7
4. Rear passengers	277	20
5. Baggage area 1 [maximum: 200 lbs]	150	14.5
6. Baggage area 2 [maximum: 50 lbs]		
7. Ramp weight and moment	2650	123.3
8. Fuel allowance for engine start and taxi	−8	−0.4
9. Take-off weight and moment (add lines 7 and 8) [maximum: 2650 lbs]	2642	122.9

Now the flight is finally safe, as well as legal!

Problem 43. Fill in the weight and balance table. Begin by loading the two baggage areas to capacity, putting all other supplies in the back seat.

	Weight (lbs)	Moment (lb.-ins./1000)
1. Basic empty weight (includes unusable fuel and full oil)	1661	63.8
2. Usable fuel (at 6 lbs/gal)[maximum: 62 gal]	264	12.5
3. Pilot and front passenger	100	3.5
4. Rear passengers	375	27.5
5. Baggage area 1 [maximum: 200 lbs]	200	19
6. Baggage area 2 [maximum: 50 lbs]	50	6
7. Ramp weight and moment	2650	132.3
8. Fuel allowance for engine start and taxi	−8	−0.4
9. Take-off weight and moment (add lines 7 and 8) [maximum: 2650 lbs]	2642	131.9

The flight is legal as far as total weight is concerned, but the center of gravity is too far back. Assume the smell of the supplies in the baggage areas can be contained, and leave the baggage areas fully loaded. Also assume that the smell of the supplies is objectionable, whether they are in the rear or the front seat. In other words, adjust the center of gravity by moving supplies from the rear seats to the (empty) front seat.

Experiment to see how many pounds must be moved from the rear to the front seat. After some trial and error, you will see that leaving 125 pounds in the rear and moving 250 pounds to the front seat results in a safe arrangement:

	Weight (lbs)	Moment (lb.-ins./1000)
1. Basic empty weight (includes unusable fuel and full oil)	1661	63.8
2. Usable fuel (at 6 lbs/gal)[maximum: 62 gal]	264	12.5
3. Pilot and front passenger	350	13
4. Rear passengers	125	9
5. Baggage area 1 [maximum: 200 lbs]	200	19
6. Baggage area 2 [maximum: 50 lbs]	50	6
7. Ramp weight and moment	2650	123.3
8. Fuel allowance for engine start and taxi	−8	−0.4
9. Take-off weight and moment (add lines 7 and 8) [maximum: 2650 lbs]	2642	122.9

Problem 44. Fill in the weight and balance table for your take-off configuration.

	Weight (lbs)	Moment (lb.-ins./1000)
1. Basic empty weight (includes unusable fuel and full oil)	1661	63.8
2. Usable fuel (at 6 lbs/gal)[maximum: 62 gal]	372	18
3. Pilot and front passenger	360	13.5
4. Rear passengers	235	17
5. Baggage area 1 [maximum: 200 lbs]	20	2
6. Baggage area 2 [maximum: 50 lbs]		
7. Ramp weight and moment	2648	114.3
8. Fuel allowance for engine start and taxi	−8	−0.4
9. Take-off weight and moment (add lines 7 and 8) [maximum: 2650 lbs]	2640	113.9

Part a. Your weight is ten pounds below the allowable maximum.

Part b, The point (113.9, 2640) lies in the envelope; therefore the plane is within legal limits for take off.

Part c. Suppose there are five gallons of fuel left on landing, and the weight of the empty picnic basket is so small that it can be ignored. Also, suppose the front and rear occupants have eaten equal amounts. (In other words, there are ten more pounds in the front seat and ten more pounds in the rear seat.) Therefore, the weight and balance table corresponding to the landing configuration looks like this:

	Weight (lbs)	Moment (lb.-ins./1000)
1. Basic empty weight (includes unusable fuel and full oil)	1661	63.8
2. Usable fuel (at 6 lbs/gal)[maximum: 62 gal]	30	2
3. Pilot and front passenger	370	14
4. Rear passengers	245	18
5. Baggage area 1 [maximum: 200 lbs]		
6. Baggage area 2 [maximum: 50 lbs]		
7. Ramp weight and moment	2306	97.8
8. Fuel allowance for engine start and taxi	−8	−0.4
9. Take-off weight and moment (add lines 7 and 8) [maximum: 2650 lbs]	2298	97.4

The point $(97.4, 2298)$ lies in the envelope, and the plane was in a safe and legal configuration when it landed.

Remark. Airplane design involves many considerations, such as the change in the center of gravity when fuel burns off during flight, and the change in the center of gravity when the landing gear retracts on take-off or extends before landing. Most planes have more than one fuel tank, and some planes have special fuel pumps to pump fuel from one tank to another, to keep the center of gravity within limits.

Haldiman——Physics; Astronaut Crew Training Instructor

Problem 45. To find each pump's speed in RPM, use the pump's gear ratio and the specified turbine shaft speed of 74,160 RPM in a fraction equation.

For the hydraulic pump, the gear ratio says that "for 18.93 spins of the turbine shaft, the hydraulic pump spins one time." So with the turbine shaft spinning 74,160 revolutions per minute, the hydraulic pump speed or spin can be calculated as follows ("h" represents the unknown hydraulic pump speed):

$$\text{gear ratio} = \text{speed ratio (i.e., spin ratio)}$$

$$18.93 : 1 = 74{,}160 : h$$

$$\frac{18.93}{1} = \frac{74{,}160}{h}$$

$$18.93 \times h = 1 \times 74{,}160$$

$$18.93 \times h = 74{,}160$$

$$h = \frac{74{,}160}{18.93}$$

$$h = 3{,}917.59 \text{ RPM}$$

So, the hydraulic pump speed is about 3,918 RPM, slower than the turbine shaft speed by a factor of 18.93.

For the lube oil pump speed, use a similar equation with the lube oil pump gear ratio of $6.07 : 1$, as follows:

$$\text{gear ratio} = \text{speed ratio (i.e. spin ratio)}$$

$$6.07 : 1 = 74{,}160 : \ell$$

$$\frac{6.07}{1} = \frac{74{,}160}{\ell}$$

$$6.07 \times \ell = 1 \times 74{,}160$$

$$6.07 \times \ell = 74{,}160$$

$$\ell = \frac{74{,}160}{6.07}$$

$$\ell = 12{,}217.46 \text{ RPM}$$

So, the lube oil pump speed is about 12,217 RPM, slower than the turbine shaft speed by a factor of 6.07, but it is not as slow as the hydraulic pump. In fact, the lube oil pump is about three times faster than the hydraulic pump.

For the APU fuel pump speed, the gear ratio is 18.93:1, the same as the hydraulic pump gear ratio. So, the APU fuel pump speed is also about 3,918 RPM, much slower than the turbine shaft speed (by the same factor of 18.93) and one-third as fast as the lube oil pump.

Problem 46. The total electrical power produced by the three fuel cells can be calculated as follows:

For fuel cell 1: 30.7 volts × 178 amps = 5464.6 watts
For fuel cell 2: 30.2 volts × 225 amps = 6795.0 watts
For fuel cell 3: 31.0 volts × 153 amps = 4743.0 watts

For total power produced, add all three amounts. The answer is 17002.6 watts, or about 17 kilowatts, or 17 KW.

Finally, since the voltage is indirectly proportional to the current, the voltage goes down when equipment is turned on, drawing electric current from a fuel cell. Fuel cell 2 has the largest current usage, and this is why it has the lowest operating voltage.

Problem 47. Since the lamp uses 75 watts, or 0.075 kilowatts (1000 watts = 1 kilowatt), and was lit every night for two hours:

$$\text{lamp KWH usage per night} = 0.075 \text{ kilowatts} \times 2 \text{ hrs per night}$$
$$= 0.150 \text{ kilowatt-hours per night}$$

For the billing period ending in May and containing 31 days, calculate:

$$\text{lamp total KWH usage for May} = 31 \text{ nights} \times 0.150 \text{ KWH per night}$$
$$= 4.65 \text{ kilowatt-hours.}$$

To convert this to a dollar amount, use the information given in the May bill to get a cost for one kilowatt-hour, then multiply this by the lamp's kilowatt-hour usage:

$$\text{cost per 1 KWH in May} = \$36.01 \text{ divided by } 445 \text{ KWH}$$
$$= \$0.0809 \text{ per 1 KWH}$$

$$\text{cost for lamp total KWH usage} = 4.65 \text{ KWH} \times \$0.0809 \text{ per KWH}$$
$$= \$0.376$$

So the total cost of lighting my bedside lamp, two hours every night, for 31 days during April/May 1990 was: 37.6 cents.

Anselm—Business Data Processing

Problem 48. **A.** To minimize read-time, use as close to 4,600 characters per block as possible. Since each record is 503 characters long, divide 4,600 by 503. Rounding to the nearest whole number the best answer is 9 records per block.

B. To minimize space, use as close to 18,400 characters per block as possible without going over. Record size is 503, so divide 18,400 by 503. A block cannot contain more than 18,400 characters, so even though the answer is more than 36.5, round down to 36 since an answer of 37 would result in 18,611 characters—more than the maximum number of characters. Using 36 records per block, find the total number of blocks needed by dividing the total number of records (115,000) by the number of records per block (36). The answer is 3,195. In this case, round up since 3,194 blocks were used, plus one partial block. This is not quite half of the disk.

C. No more than 6,800 blocks can be used, and there are 115,000 records to store using as close to 4,600 characters per block as possible. One way to figure the answer is to divide the 115,000 records by the total number of blocks (6,800). Taking the answer 16.91, round up to 17. (Sixteen is closer to the most desirable answer of nine, but 16 would require more than 6,800 blocks). For this question the best answer is 17 records per block.

Note that by using a block size for optimal read-and-write-time, 115,000/9 or 12,778 total blocks would be used—nearly two magnetic disks. Minimizing space uses 3,194 blocks, which is about 47% of one disk. Very often in business data processing there are compromises, as in example C. The goal is to spend as little money as possible on equipment (in this example no more than one magnetic disk) while reading and writing the file as efficiently as possible.

Problem 49. The coded message is, "THIS IS HOW A COMPUTER WORKS." At address 10 is a "T", the first letter of the message. At address 11 is a "1", and at address 12 is a "6". This directs you to address 16, which contains an "H". The next address 85 contains an "I". Following the chain through should bring you to address 38, which contains an "S". You can determine that this is the last letter by looking at the next address, which is pointed to at addresses 39 and 40. Each contained an "F", indicating the end of the message. This is very similar to how a computer stores information in its memory. Most computers, however, store groups of characters chained together, not single characters as in this example.

Problem 50. This solution involves "Thinking Outside the Box." Most people approach this problem by dividing the shape into three equal squares or six triangles, because they are so commonly used. However, the actual solution is shown in Figure 71.

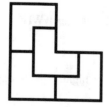

FIGURE 71

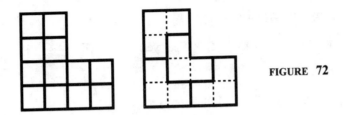

FIGURE 72

One way to come up with the solution is to divide the shape into twelve equal boxes, and then assemble the twelve boxes into four groups of three.

Agmon—Software Engineering; Real Estate Investment

Problem 51.   Total yearly debts allowed:

$$40\% * \$65,000 = \$26,000 \text{ per year}$$

Mortgage loan amount:

$$10\% * \text{mortgage} = \$26,000$$

$$\text{mortgage} = \$26,000/.10 = \$260,000$$

Problem 52.   House price:

$$80\% * \text{price} = \text{mortgage} = \$260,000$$

$$\text{price} = \$260,000/.80 = \$325,000$$

$$\text{down payment} = .20 * \$325,000 = \$65,000$$

Problem 53.   Car loan yearly debt:

$$\$500 * 12 = \$6000 \text{ per year}$$

Student loan yearly debt:

$$\$200 * 12 = \$2400 \text{ per year}$$

Yearly debt for mortgage loan:

$$\$26,000 - \$6000 - \$2400 = \$17,600 \text{ per year}$$

Mortgage loan amount * 10% interest = $17,600 per year

$$\text{mortgage loan} = \$17,600/.10 = \$176,000$$

House price * 80% = $176,000

$$\text{price} = \$176,000/.80 = \$220,000$$

Problem 54. Rental property mortgage debt:

$$\$100,000 * 8\% = \$8000 \text{ per year}$$

Rental income:

$$\$1,250 * 12 = \$15,000 \text{ per year}$$

Yearly debt for mortgage loan:

$$\$17,600 - \$8000 + \$15,000 = \$24,600 \text{ per year}$$

Mortgage loan amount $* 10\%$ interest $= \$24,600$

$$\text{loan amount} = \$24,600/.10 = \$246,000$$

House price $* 80\% = \$246,000$

$$\text{price} = \$246,000/.80 = \$307,500$$

Eckerle—Quality Engineering

Problem 55.

$$CR = \frac{6s}{\text{tolerance}} = \frac{6(.25\text{mm})}{2\text{mm}} = \frac{1.5\text{mm}}{2\text{mm}} = .75$$

Change to a percent $.75 * 100 = 75\%$

$$CP = \frac{\text{tolerance}}{6s} = \frac{2\text{mm}}{6(.25)} = \frac{2\text{mm}}{1.5\text{mm}} = 1.33$$

$$CPK = \text{min of}$$

$$\frac{(USL - \bar{x})}{3s} = \frac{(16\text{mm} - 15.3\text{mm})}{3(.25\text{mm})} = \frac{.7\text{mm}}{.75\text{mm}} = .933$$

$$\frac{(\bar{x} - LSL)}{3s} = \frac{(15.3\text{mm} - 14\text{mm})}{3(.25\text{mm})} = \frac{1.3\text{mm}}{.75\text{mm}} = 1.733$$

so $CPK = \min(.933, 1.733) = .933$.

Control Limits for Center:

$$UCL = \bar{\bar{x}} + A_2\bar{r} = 15.3\text{mm} + (1.023)(.15\text{mm}) = 15.45\text{mm}$$

$$LCL = \bar{\bar{x}} - A_2\bar{r} = 15.3\text{mm} - (1.023)(.15\text{mm}) = 15.147\text{mm}.$$

Control Limits for Ranges:

$$UCL = D_4\bar{r} = 2.574(.15\text{mm}) = .3861\text{mm}$$

$$LCL = D_3\bar{r} = 0(.15\text{mm}) = 0\text{mm}$$

In this case, the Quality Engineer placed the points on the chart and, based on the definition for stability, found the process to be stable and predictable over time.

The CP and CR were okay, but the CPK showed that the process was off-target. It was centering on the high side of the specifications, so parts could have been made that would be unacceptable to the customer. The Quality Engineer told the operator to adjust the machine, bringing the part sizes on-target.

Problem 56.

$$\bar{x} = (65 + 35 + 40 + 30 + 50 + 70 + 25 + 55 + 40 + 60)/10$$

$$= 47 \text{ minutes}$$

Janine was surprised that on average, she spent less than an hour per day on homework. She also noticed that on some days, she spent more time than on others. She decided to investigate the reasons for the differences, so she could even out the times.

Problem 57.

training	$150 \times \$30 =$	\$ 4500
design review	$40 \times \$35 =$	1400
audits	$20 \times \$25 =$	500
process control	$250 \times \$25 =$	6250
incoming inspection	$50 \times \$25 =$	1250
scrap		55000
rework		70000
warranty		+35000
Total		\$173,900

Which category of quality costs was the largest?

$$\text{Prevention} = \$\ \ 6,400$$
$$\text{Appraisal} = \ \ 7,500$$
$$\text{Failure} = \ \ 160,000$$

Failure costs were the largest. To reduce those costs, the Apez Company needs to work on the failure causes.

Lipsey—Health Science

Problem 58. To find the patient's weight in kg, multiply by the conversion ratio 1 kg/2.2 lb. Thus 156 lb ∗ 1 kg /2.2 lb =70.9 kg.

To find the amount in mg to be injected, multiply the amount per kg by the number of kilograms, and round to 2 decimal places. Thus, $0.005 * 70.9 = 0.35$ mg.

Problem 59. To find the percent of total beds devoted to critical care, divide the number of critical care beds by the total number of beds, multiply by 100%, and round to 1 decimal place. For the larger hospital, $30/360*100\% = 8.3\%$; for the smaller hospital, $14/164*100\% = 8.5\%$. Since $8.5\% > 8.3\%$, the smaller hospital has a greater percentage of critical care beds.

Problem 60. First find T/5. Since T = 80, T/5 = 80/5 = 16. Substitute the appropriate values for C, H, and T/5:

$$C - H - T/5 = 230 - 50 - 16 = 180 - 16 = 164$$

Thus LDL = 164 > 160. The patient is in the high-risk category.

Bowen—Nursing Education

Problem 61. There is a two-step solution: If Mr. Jones needs 125 ml each hour, then 125 divided by 60 minutes equals the number of ml each minute (2.08 ml). If each ml requires 15 drops, then 15 times 2.08 ml gives the rate of drops for each minute (31.2 or 31 when rounded off). The nurse counts the drops and regulates the rate to 31 drops per minute using a roller clamp device on the tubing to keep the drip rate constant.

Problem 62. This requires four steps to solve. First, convert liters to milliliters, because the tubing information uses this form. One liter equals 1000 ml. Second, divide the total volume (1000 ml) by the number of hours (10) to get the volume for each hour (100 ml). The third and fourth steps are the same steps as suggested above.

Problem 63. Since Paul's weight is given in kilograms, the first step is to multiply that weight times 20 mg for the total number of milligrams needed. In this case the safe dose is 260 mg.

Second, one must find out how much liquid in milliliters to give Paul, because this medication comes as a liquid. There are 250 mg per 5 ml. Therefore, Paul needs just a little over five ml.

Medications dispensed for home use are usually dispensed in household measurements: for example, Paul would get one teaspoon of amoxicillin per dose.

Problem 64. This solution involves several steps.

First, convert Mandy's weight from pounds to kilograms. There are 2.2 pounds to a kilogram. Forty divided by 2.2 equals 18.1 kg.

Second, determine how much medication is safe for that weight. If 40 mcg are safe for one kg, then 40 times 18.1 or 724 mcg are safe for Mandy.

Third, this medication comes as a liquid with 50 mcg in each ml. Since Mandy's dose is 724 mcg, this amount is divided by 50 mcg for the total number of milliliters (ml) needed. 724 divided by 50 yields 14.48 ml needed for Mandy's dose of digoxin.

This dose would be measured in a syringe marked in tenths of milliliters for accuracy, rounding off to the next lowest tenth (14.4 ml) to avoid overdosing the child.

Problem 65. The formula for this type of problem is stated as drug ordered divided by drug on hand times volume on hand:

$$\frac{3 \text{ mg}}{10 \text{ mg}} \times 2 \text{ ml} = \frac{6}{10} \text{ml} \quad \text{(usually written as 0.6 ml).}$$

The multiplication of fractions gives the answer: 0.6 ml.

Problem 66. This problem uses two measurement systems: the metric system (ml) and the apothecary system (grains). Since the drug comes supplied in mg, it is best to convert the grains to mg and then solve the problem as in #3. There are 60 mg in one grain.

Therefore, $\frac{1}{10}$ grain is $\frac{1}{10}$ of 60 mg or 6 mg. To continue,

$$\frac{6 \text{ mg}}{10 \text{ mg}} \times 1 \text{ ml} = \frac{6}{10} \text{ ml} \quad (0.6 \text{ ml}).$$

Problem 67. Simply convert grains to milligrams. If one grain = 60 milligrams, then 0.5 grain = 30 milligrams. Thirty milligrams is certainly more than 15 milligrams, so her dose was decreased significantly. The elderly frequently need lower doses of medications.

Problem 68. Convert in either direction: micrograms to milligrams or the other way around. One milligram has 1000 micrograms, so 0.3 milligrams has $\frac{3}{10} \times 1000$ or 300 micrograms (mcg). There has been no error.

Gerson—Electrical Engineering; Space Systems

Problem 69. Part a. See the graph in Figure 73.

    Part b. t = time in minutes

           I_B = battery discharge in amperes

$$I_B(t) = \begin{cases} 10 \sin \dfrac{\pi t}{15} & 0 \le t \le 15 \\ 0 & 15 \le t \le 60 \\ 30 & 60 \le t \le 90 \end{cases}$$

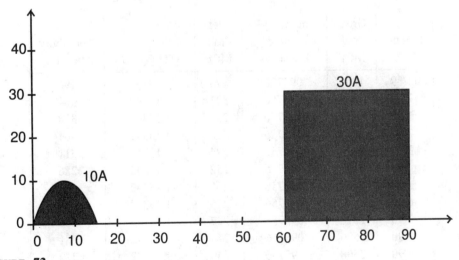

FIGURE 73
Battery Discharge vs. Time graph

Part c. I'_{BC} = cumulative battery discharge in ampere-minutes
I_{BC} = cumulative battery discharge in ampere-hours

$$I'_{BC} = \int_0^{15} 10 \sin \frac{\pi t}{15} \, dt + \int_{15}^{60} 0 \, dt + \int_{60}^{90} 30 \, dt$$

$$= 10 \left[\left(\frac{-15}{\pi} \cos \frac{\pi t}{15} \right) \Big|_0^{15} \right] + 0 + (30t) \Big|_{60}^{90}$$

$$= 10 \left[\frac{-15}{\pi} \cos \pi + \frac{15}{\pi} \cos 0 \right] + [30(90) - 30(60)]$$

$$= 10 \left[\frac{15}{\pi} + \frac{15}{\pi} \right] + 900$$

$$= \frac{300}{\pi} + 900 = 995.5 \text{ ampere-minutes}$$

$$I_{BC} = \frac{I'_{BC}}{60} = 16.6 \text{ ampere-hours.}$$

Part d. C_B = battery capacity in ampere-hours

$$\frac{I_{BC}}{C_B}(100\%) = \frac{16.6}{50}(100\%) = 33\%$$

Therefore the cumulative battery discharge constraint is not violated.

Halpin—Oil and Gas Accounting

Problem 70. If you have access to a computer and spreadsheet software, try using them to solve this problem.

Month/ Year	Gross Prod. MCF	Thornbury's Share of Revenues	Net Prod. MCF	BTU @14.73	Thornbury Allocation
11/89	1806	.095756	173	1.13342	196
12/89	3354	.095756	321	1.13342	364
1/90	3112	.095756	298	1.13342	338
2/90	2777	.095756	266	1.13342	301
3/90	2858	.095756	274	1.13342	311
4/90	2629	.095756	252	1.13342	286
5/90	2769	.095756	265	1.13342	300
6/90	2567	.095756	246	1.13342	279
7/90	2578	.095756	247	1.13342	280
8/90	2590	.095756	248	1.13342	281
9/90	2508	.095756	240	1.13342	272
10/90	2494	.095756	239	1.12632	269

Month/ Year	Thornbury Allocation	Price	Earned Income	Recoup 75%	Receive 25%
11/89	196	$3.00	$588	$441	$147
12/89	364	$3.00	$1092	$819	$273
1/90	338	$3.00	$1014	$761	$253
2/90	301	$3.00	$903	$677	$226
3/90	311	$3.00	$933	$700	$233
4/90	286	$3.00	$858	$644	$214
5/90	300	$3.00	$900	$675	$225
6/90	279	$3.00	$837	$628	$209
7/90	280	$3.00	$840	$630	$210
8/90	281	$3.00	$843	$632	$211
9/90	272	$3.00	$816	$612	$204
10/90	269	$3.00	$807	$605	$202
Totals			$10,431	$7,824	$2,607

Column calculations:

$$\text{Gross Production MCF} * \text{Thornbury's Share} = \text{Net Production MCF}$$

$$\text{Net Production MCF} * \text{BTU rate} = \text{Thornbury's Allocation}$$

$$\text{Allocation} * \text{Price} = \text{Earned Income}$$

$$\text{Recoupment} = 75\% \text{ of Earned Income}$$

$$\text{Receive} = 25\% \text{ of Earned Income}$$

Part a. The total earned income for Mr. Thornbury during this period is $10,431.

Part b. The amount recouped by ARKTX is $7,824.

Part c. Mr. Thornbury has actually received $2,607.

Part d. Mr. Thornbury still owes ARKTX $30,673.48 − $7,824 = $22,849.48

Leva—Business Administration Higher Education

Problem 71. As with most computer programs, large or small, there are many correct answers. Here is one solution, using the computer language FORTRAN. If your solution is different, it is still correct if it yields the right answers for the test inputs.

Step 1. Change the recorded price, which is a real number, to an integer, eliminating the incorrect decimal portion.

```
IPRICE = IFIX (DATA)
```

If DATA was read in as 12.7, the new integer value stored in IPRICE is simply 12, with no decimal point or decimal portion.

Step 2. Restate new integer value as a real number.

```
REPRIC = FLOAT (IPRICE)
```

The new real value stored in REPRIC is 12.0.

Step 3. Separate the incorrect decimal portion of the original value from the dollar portion.

X = DATA - REPRIC

Because DATA equals 12.7 and REPRIC equals 12.0, the new value of X will be 0.7.

Step 4. Calculate the actual stock price.

PRICE = X / .8 + REPRIC

This calculation adds the fraction 7/8 to 12.0 yielding 12.875.

Formatted Price	Decimal Price
29.0	29.0
30.1	30.125
10.2	10.25
15.3	15.375
112.4	112.5
20.5	20.625
50.6	50.75
33.7	33.875

A formatted stock price of 11.8 or 15.9 would be an error because in this notation, the number after the decimal place represents eighths, not tenths, so $11.8 = 11 + 8/8 = 11 + 1 = 12$, so instead of storing 11.8, the formatted stock price would be 12.0. Similarly there would never be a formatted stock price of 15.9 because $9/8 = 11/8$. Instead of storing 15.9, the formatted stock price would be stored as 16.1. Using the formatted stock prices, the decimal place is always exactly one digit, and that digit must be between 0 and 7.

Nguyen—Aerospace Engineering

Problem 72. For MEO orbit: Four batteries are needed.
For GEO orbit: Four batteries are needed.

Lanham—Structural Engineering

Problem 73. Step 1.

$$P = 100{,}000 \text{ lbs}$$

$$F_c = 2450 \text{ lbs/in}^2$$

$$F_c = \frac{P}{A}$$

$$2450 \text{ lbs/in}^2 = \frac{100{,}000 \text{ lbs}}{(12 \text{ in})(x \text{ in})}$$

$$x = \frac{100{,}000 \text{ lbs}}{(12 \text{ in})(2450 \text{ lbs/in}^2)} = 3.40 \text{ in}$$

The unknown width of the post is 3.40 inches.

Step 2. To convert the allowable bearing pressure from tons/ft^2 to lbs/in^2:

$$\left(\frac{4\text{ tons}}{\text{ft}^2}\right)\left(\frac{2000\text{ lbs}}{\text{ton}}\right)\left(\frac{1\text{ ft}^2}{144\text{ in}^2}\right) = 55.6\text{ lbs/in}^2$$

$$P = 100{,}000\text{ lbs}$$

$$F_c = 55.6\text{ lbs/in}^2$$

$$F_c = \frac{P}{A}$$

$$55.6\text{ lbs/in}^2 = \frac{100{,}000\text{ lbs}}{(12\text{ in})(y\text{ in})}$$

$$y = \frac{100{,}000\text{ lbs}}{(12\text{ in})(55.6\text{ lbs/in}^2)} = 150\text{ in}$$

The bearing plate is 12 inches by 150 inches, so the required area for the bearing plate resting on gravel is 1800 in^2. A square bearing plate would be 42.5 in. wide since $\sqrt{1800\text{ in}^2} = 42.5$ in.

Step 3.

$$P = 2450\text{ lbs}$$

$$A = 0.5\text{ in}^2$$

$$F_{\text{shear}} = \frac{P}{A} = \frac{2450\text{ lbs}}{0.5\text{ in}^2} = 4900.0\text{ lbs/in}^2$$

Since the ultimate shearing stress of structural steel is 25,000 lbs/in^2, the 0.5-inch plate is adequate. A thinner plate or a bearing plate made of cheaper material could be used instead.

Parker—Computer Science

Problem 74. Here is some sample input data:

```
F-number: F123-874
Customer: Mara Salem
City-State: Frisky Kittens, UT
Zip: 80315
Phone: 801-555-3434
Comments: Had questions about product
F-number: F123-899
Customer: Janie Moose
Address: 2435 Dindar
City-State: Austin, TX
Fax: 505-555-3435
F-number: F123-222
```

```
Customer: Amber MacLyle
Address: 12567 Franklin Way
City-State: Boulder, CO
Zip: 80306
Phone: 303-555-1234
Fax: 505-555-3435
```

The script would produce this output from the above input:

```
F123-874
Mara Salem

Frisky Kittens, UT
80315
801-555-3434
F123-899
Janie Moose
2435 Dindar
Austin, TX
80315
801-555-3434
505-555-3435
```

Problem 75. The bugs are explained in problems 76 and 77, but they are illustrated by the example in the answer to Problem 74. Note that the phone number and zip code for Mara Salem are printed out for Janie Moose. There should be blank lines for the phone and zip fields for Janie Moose, since that data was missing from the input. Also note that the data for Amber MacLyle does not appear in the output at all.

Problem 76. The first bug, which caused Mara Salem's phone number to be printed out again for Janie Moose, can be fixed by rewriting the code as follows (some of the comments have been removed to make it more readable):

```
    BEGIN code
# this code will happen only once, before reading any of the input
    set first = 1
    set FS = '':"
    EACH LINE code
# this code will happen once for every line of input
    if ($1 = ''F-number" ) then
    if (first = 1) then
    set first = 0
    else
# Output the completed record for the previous fnum
# This is the only place in the script that produces output
```

```
        print fnum
        print customer
        print address
        print citystate
        print zip
        print phone
        print fax
# here is the first bug fix
# moved these lines outside of the if-then-else statement that
# begins with if (first = 1) then ... else ... so that they will
# happen every time we are inside the if ($1 = ''F-number") then
# regardless of whether or not first = 1
#
# initialize the fields to empty strings so if an
# input field is missing, a blank line will be printed in
# the output in place of the missing field
        set customer = " "
        set address = " "
        set citystate = " "
        set zip = " "
        set phone = " "
        set fax = " "
        set fnum = $2
        if ($1 = ''Customer") then
        set customer = $2
        if ($1 = ''Address" ) then
        set address = $2
        if ($1 = ''City-State") then
        set citystate = $2
        if ($1 = ''Zip") then
        set zip = $2
        if ($1 = ''Phone") then
        set phone = $2
        if ($1 = ''Fax") then
        set fax = $2
# note if $1 = anything else, the input line will simply be ignored
        END code
# this code happens once, after the last line of input
# nothing special to do here...
```

By moving the initializations to blank (" ") outside of the "if first..." statement, they will happen for every new F-number. It is important to initialize the variables to blank after doing the prints of the previous F-number data. If you reset the variables to blank before doing the prints of the previous F-number data, you would never print out anything other than blanks.

Problem 77. The second bug can be fixed by adding print statements to the END section, as follows:

```
    END code
# this code happens once, after the last line of input has gone
# print out the last set of data
    print fnum
    print customer
    print address
    print citystate
    print zip
    print phone
    print fax
```

So every line of output, except the last line, will be printed out in the EACH LINE code, while processing the next line of input. The new code we have added to the END code will print out the last line of output at the very end, which happens exactly once after processing all the input.

You can create more input and run it through the program with both bug fixes to see if it generates the right output. It is entirely possible there are other bugs, and that there are other, better ways to write this program.

Problem 78. The first plane in Figure 36 is heading 230. Figure 74 demonstrates that with the heading indicator showing a present heading of 230, the 090 radial falls into area B, indicating the direct entry shown in Figure 36.

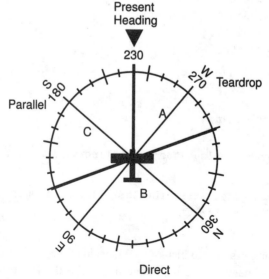

FIGURE 74
Heading indicator for first plane in Figure 36

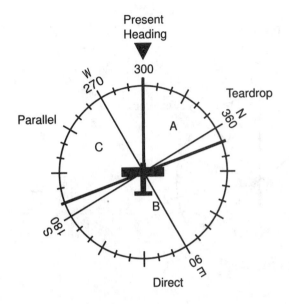

FIGURE 75
Heading indicator for second plane in Figure 36

Similarly, the heading of the second plane is 300. It also should make a direct entry.

The heading of the plane in Figure 37 is 070. Figure 76 shows why the entry from the fix to the hold on radial 090 should be teardrop: the 090 radial falls into area A.

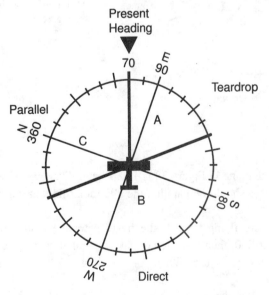

FIGURE 76
Heading indicator for plane in Figure 37

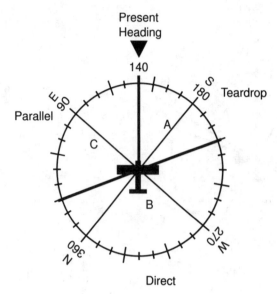

FIGURE 77
Heading indicator for plane in Figure 38

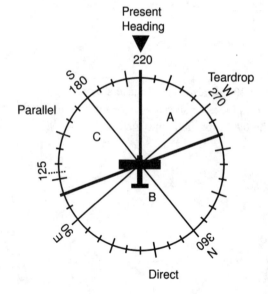

FIGURE 78
Heading 220, radial 125 shown for hold
in Figure 33

The heading of the plane in Figure 38 is 140. Figure 77 shows why the entry from the fix to the hold on radial 090 should be parallel: the 090 radial falls into area C.

Problem 79. If you are flying towards the fix to enter the hold shown in Figure 33, and your present heading is 220, then you should make a parallel entry. The radial that defines the holding pattern in Figure 33 is 125. When your heading is 220, the 125 radial falls into area C, for a parallel entry.

If your heading is 030, the entry should be direct. If your heading is 100, the entry should be teardrop.

Problem 80. The radial that defines the hold in Figure 34 is 310. If the present heading is 180, then the 310 radial will fall into area B, so the entry would be direct.

If the heading is 300, then the 310 radial will fall into area A, so the entry would be teardrop.

If the present heading is 360, then the 310 radial will fall into area C, so the entry should be parallel.

Problem 81. The radial that defines the non-standard, all-left turns holding pattern in Figure 35 is 200. If the present heading is 250, then the 200 radial falls into area A, so the entry is teardrop.

If the present heading is 120, the 200 radial would fall in area C, so the correct entry should be parallel.

If the present heading is 020, then the 200 radial would fall in area B, so the correct entry should be direct.

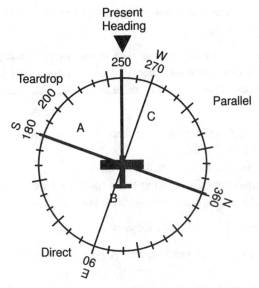

FIGURE 79
Heading 250, radial 200 shown for hold in Figure 35

Poiani—Mathematics

Problem 82. Let us begin with some terminology. The diagram is called a *network*. Point I is a *vertex,* and so are N, S, and E. The interconnections of these *vertices* are called *arcs.* The key to the solution is the "order of the vertices", or the number of arcs emanating from a given vertex. This number can be odd or even (called *odd* or *even vertex*).

A *path* in a network is a series of arcs that can be traveled continuously without retracing any arc. A path *traverses* the network if every arc of the network is covered by the path. A path is *closed* if the starting and ending vertices are the same. These networks are connected.

Consider the child's game:

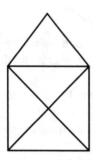

FIGURE 80
Child's Game

Can you traverse this network with one closed path? The answer is "no" because not all the vertices are even.

Theorem. *A network can be traversed by one path beginning and ending at the same vertex if (and only if) all of the vertices are even. That is because even vertices pair one exit-arc with one entrance-arc, so you can go in and out of them freely, beginning and ending at the same vertex.*

So, how many even and/or odd vertices are there in the Königsberg Bridge network? Can it be traversed by a single closed path?

The answer is "no" because all four vertices are odd.

To figure out how many separate walking tours (paths) would be necessary to pass over each of the seven bridges once, we must analyze these odd vertices.

Theorem. *The number of odd vertices in a network is always even.*

Proof. Let V be the number of odd vertices, and A be the number of arc ends. Note that A must be an even number, since every arc has exactly two ends.

If v_i is the number of vertices of order i, then:

$$V = v_1 + v_3 + v_5 \cdots + v_{2k+1}$$

$$A = v_1 + 2v_2 + 3v_3 + 4v_4 \cdots + rv_r$$

So,

$$A - V = (v_1 + 2v_2 + 3v_3 + 4v_4 \cdots) - (v_1 + v_3 + v_5 \cdots)$$

$$= (v_1 - v_1) + 2v_2 + (3v_3 - v_3) + 4v_4 + (5v_5 - v_5) + 6v_6 + \cdots$$

$$= 0 + 2v_2 + 2v_3 + 4v_4 + 4v_5 + 6v_6 + 6v_7 + \cdots$$

$$= 2(v_2 + v_3) + 4(v_4 + v_5) + 6(v_6 + v_7) + \cdots + 2n(v_n + v_{n+1})$$

Thus, $A - V$ is an even number. Since we know A is an even number, V must also be an even number.

Theorem. *If a network has 2n, n = 1, 2, 3, ..., odd vertices, then it can be traversed by n paths whose initial and terminal vertices are odd.*

This is stated without formal proof. Just think how the oddness leaves you with "extra arcs" that need to be reached through extra paths.

For the child's game (vertex order is in parentheses):

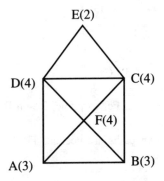

FIGURE 81
Child's game with labeled vertices

The two odd vertices are *A* and *B*. Thus, the single path must begin at *A* and end at *B* (or vice versa).

For the Königsberg Bridge problem: All four vertices are odd. By Theorem C, the network has $4 = 2 * 2 = 2n$ odd vertices, so it can be traversed by $n = 2$ paths whose initial and terminal vertices are the odd ones. Thus, you need two separate paths to traverse the network.

Problem 83. The "order" of a vertex is the number of arc-edges connected to that vertex. Figure 82 shows the order of each vertex in parentheses.

From the order of the vertices, two are odd. Thus, the network requires traversing one path, that begins and ends at the odd vertices *A* and *E*. Try tracing out the path. There is more than one possible route.

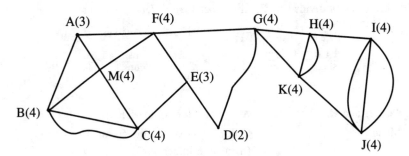

FIGURE 82
Network of roadways with vertex orders

Problem 84. To solve this problem, represent the house as a graph. Each room becomes a vertex, and each door becomes an edge. Just as a door connects two rooms, an edge connects

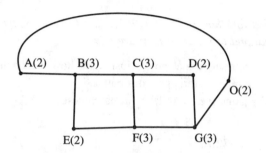

FIGURE 83
House floor plan as a graph

two vertices. The order of each vertex is shown in Figure 83. For example, Room B has three doors, or three edges in this graph, so its order is three.

There are four odd vertices in this floorplan, so two paths are needed. The paths begin and end at odd vertices. Try tracing some interesting routes.

Siler—Dietetics

Problem 85. **A.** 46 divided by 16 = 2.875 or 2 7/8 lbs.

B. 3/4 cup * 8 oz/cup * 2 Tbsp/oz = 12 Tbsp.

C. 1/2 gal = 8 cups.

D. 3 oz * 2 Tbsp/oz * 3 tsp/Tbsp = 18 tsp.

E. 6.5 qts * 32 oz per qt = 208 oz.

F. 2 gal * 4 qts/gal * 2pts/qt = 16 pts

G. 2.5 gal * 4 qts per gal * #24 dipper. Answer: c. 240

H. 32 oz per qt. Answer: d. #30

I. 4 gal * 4 qts per gal * #20 dipper. Answer: a. 320

J. 200 bread slices divided by 28 slices per loaf = 7.14. Answer: b. 7

K. 200 slices $* \dfrac{1.5 \text{ tsp}}{\text{slice}} * \dfrac{\text{Tbsp}}{3 \text{ tsp}} * \dfrac{\text{lb}}{32 \text{ Tbsp}} = 3.125$ lbs. Answer: d. 3.0

L. 1000 slices $* \dfrac{1 \text{ tsp}}{\text{slice}} * \dfrac{1 \text{ Tbsp}}{3 \text{ tsp}} * \dfrac{1 \text{ oz}}{2 \text{ Tbsp}} * \dfrac{1 \text{ cup}}{8 \text{ oz}} * \dfrac{1 \text{ pt}}{2 \text{ cups}} * \dfrac{1 \text{ gal}}{8 \text{ pts}} = 1.3$ gal. Answer:
d. 1 1/3 gal

Problem 86. **A and B.** (Write down your menu for comparison.)

Sample Menu		Guide To Good Eating Comparison (measured in servings)
Breakfast		
Cornflakes	1 cup	1 Grain-Starch
Milk	4 oz	1/2 Milk
Banana	small	1 Fruit-Vegetable
Orange Juice	1 cup	2 Fruit-Vegetable

Lunch

Cheeseburger	1/4 lb	2 Grain-Starch
		2 Meat
		1/2 Milk
with lettuce & tomato		1/2 Vegetable
Diet Coke	16 oz	1 Other Food

Supper

Baked Chicken Breast	1	2 Meat
Baked Potato	1	1 Fruit-Vegetable
with Butter	2 tsp	1 Other Food
with Cheese	1 oz	1/2 Meat or 1/2 Milk
Green Beans	1/2 cup	1 Fruit-Vegetable
Fresh Fruit Salad	1/2 cup	1 Fruit-Vegetable
Hard Roll	1	1 Grain-Starch
Butter	1 tsp	1 Other Food
Carrot Cake	1 slice	1 Other Food

Totals

Milk Group	1 1/2 (Short 1/2 serving for the day)
Meat Group	4
Fruit-Vegetable Group	6 1/2
Grain-Starch Group	4

Problem 87. A. Jim:

IBW: 184 lbs

Calorie requirements:

$$BEE = 66 + (13.7 * 184/2.2) + (5 * 73 * 2.54) - (6.8 * 35) = 1900$$

Calories = BEE * activity factor = $1900 * 1.7 = 3230$

Protein requirements: 67–84 grams

Fluid requirements: 2943 cc = 12 1/4 8 oz glasses of water

B. Jane

IBW: 125

Calorie requirements:

$$BEE = 655 + (9.6 * 125/2.2) + (1.7 * 65 * 2.54) - (4.7 * 30) = 1340$$

Calories = BEE * activity factor = $1340 * 1.3 = 1742$

Protein requirements: 49–61 grams

Fluid requirements: 2147 cc = 9 8 oz glasses of water

C. Sara

IBW: 135 lbs

Calorie requirements:

$$BEE = 655 + (9.6 * 165/2.2) + (1.7 * 67 * 2.54) - (4.7 * 22) = 1561$$

Calories = BEE * activity factor = $1561 * 1.3 = 2029$

Protein requirements: 60–75 grams

Fluid requirements: 2625 cc = 11 8 oz glasses of water

Problem 88.   **A.** IBW: 125 lbs, %IBW: 102, %UBW: 88
  B. IBW: 178, %IBW: 84, %UBW: 88

Problem 89.   **A.** 27-yr-old male:
  IBW: 172
  BEE $= 66 + (13.7 * 200/2.2) + (5 * 71" * 2.54) - (6.8 * 27) = 2029$
  Calories $=$ BEE $*$ activity factor $= 2029 * 1.3 = 2638$
  Protein: 73-91 grams
  Fluid: 3185 cc or 13 1/4 cups

  B. 33-yr-old female:
  IBW: 115
  BEE $= 655 + (9.6 * 130/2.2) + (1.7 * 63 * 2.54) - (4.7 * 33) = 1339$
  Calories $=$ BEE $*$ activity factor $= 1339 * 1.3 = 1741$
  Protein: 47-59 grams
  Fluid: 2068 cc or 69 oz or 8 2/3 cups

  C. 81-yr-old male:
  IBW: 166
  BEE $= 66 + (13.7 * 180/2.2) + (5 * 70 * 2.54) - (6.8 * 81) = 1525$
  Calories $=$ BEE $*$ activity factor $= 1525 * 1.3 = 1983$
  Protein: 66-82 grams
  Fluid: 2460 cc or 10 1/4 cups

  D. 83-yr-old female:
  IBW: 110
  BEE $= 655 + (9.6 * 110/2.2) + (1.7 * 62 * 2.54) - (4.7 * 83) = 1013$
  Calories $=$ BEE $*$ activity factor $= 1013 * 1.3 = 1317$
  Protein: 40-50 grams
  Fluid: 1500 cc or 6 1/4 cups

Problem 90.   IBW: 135 lbs
  %IBW $= 111\%$ (over 100% is considered overweight)
  %UBW $= 86\%$

Problem 91.   **A.** 120 lbs
  B. 1374
  C. 1924
  D. 1424
  E. Protein 53 grams, carbohydrate 213 grams, fat 40 grams
  F. 474 calories, 71 grams carbohydrate

Problem 92.

	Breakfast	Lunch	Supper	Carbohydrate	Protein	Fat
Lean meat	1	3	2	0	42	18
Starch	2	2	2	90	18	0
Vegetable	—	1	2	15	6	0
Fruit	1	1	1	45	0	0
Fat	1	2	2	0	0	25
Lowfat milk	1	1/2	1/2	24	16	10
			Totals:	174	82	53

Total calories: 1501

Problem 93. **A.** (This is not the only correct answer.)

275 grams carbohydrate: 55 percent

100 grams protein: 20 percent

56 grams fat: 25 percent

B. One possible answer:

	Breakfast	Lunch	Supper	Snack	Carbohydrate	Protein	Fat
Lean meat	1	2	2	1	0	42	18
Starch	3	2	3	2	150	30	0
Vegetable	0	2	2	0	20	8	0
Fruit	1	1	1	0	45	0	0
Fat	2	2	2	1	0	0	35
Skim milk	1	1	1	0	36	24	0
				Totals:	251	104	53

Total calories: 1897

Problem 94. **A.** Diet order: 100 ML Osmolite every hour

Total ml in 24 hours: 100 ml $*$ 24 hours = 2400 ml

Total calories: 2400 ml $*$ 1.06 calories per ml = 2544

Protein (in grams): 37.2 $*$ 2400 ml divided by 1000 = 89.28 grams

USRDA vitamins and minerals (percentage): 100 $*$ 2400 ml divided by 1887 ml
= 127.2%

Water from formula (in cc): 2400 ml $*$ .85% = 2040 cc

Grams of nitrogen: 89.28 divided by 6.25 = 14.29

Non protein kilocalories-to-nitrogen ratio:

89.28 $*$ 4 = 357 protein calories

2544 − 357 = 2187 non protein calories

$$\frac{\text{non protein kcalories}}{\text{nitrogen}} = \frac{2187}{14.29} = 153$$

The total kilocalories-to-nitrogen ratio:

$$\frac{\text{total kcalories}}{\text{nitrogen}} = \frac{2544}{14.29} = 178$$

B. Diet order: Jevity 1600 ml every 24 hours

    Total ml in 24 hours: 1600 ml

    Total calories: 1600 ml * 1.06 = 1696

    Protein (in grams): 44 gms * 1600 ml divided by 1000 = 70.4 grams

    USRDA vitamins and minerals (percentage): 100 * 1600 ml divided by 1321 ml
       = 121%

    Water from formula (in cc): 1600 ml * .85% = 1360 ml

    Grams of nitrogen: 70.4 divided by 6.25 = 11.264

    Non protein kilocalories-to-nitrogen ratio:

       70.4 * 4 = 281 protein calories

       1696 − 281= 1415 non protein calories

$$\frac{\text{non protein kcalories}}{\text{nitrogen}} = \frac{1415}{11.264} = 126$$

    The total kilocalories-to-nitrogen ratio:

$$\frac{\text{total kcalories}}{\text{nitrogen}} = \frac{1696}{11.264} = 151$$

C. Diet order: Enrich 240 l four times a day

    Total ml in 24 hours: 240 ml * 4 = 960 ml

    Total calories: 960 ml * 1.1 calories = 1056

    Protein (in grams): 39.7 * 960 divided by 1000 = 38.1 grams

    USRDA vitamins and minerals (percentage): 100 * 960 ml divided by 1391 = 69%

    Water from formula (in cc): 960 ml * .85 = 816 cc

    Grams of nitrogen: 38.1 divided by 6.25 = 6.1

    Non protein kilocalories-to-nitrogen ratio:

       38.1 * 4 = 152.4 protein calories

       1056 − 152 = 904 non protein calories

$$\frac{\text{non protein kcalories}}{\text{nitrogen}} = \frac{904}{6.1} = 148.2$$

    The total kilocalories-to-nitrogen ratio:

$$\frac{\text{total kcalories}}{\text{nitrogen}} = \frac{1056}{6.1} = 173$$

D. Diet order: Traumacal 50 ml every hour

    Total ml in 24 hours: 50 ml * 24 hours = 1200 ml

    Total calories: 1200 ml * 1.5 = 1800

    Protein (in grams): 83 grams * 1200 divided by 1000 = 99.6 grams

    USRDA vitamins and minerals (percentage): 100 * 1200 ml divided by 2000 ml
       = 60%

    Water from formula (in cc): 1200 ml * 77.6 = 931 ml

    Grams of nitrogen: 99.6 divided by 6.25 = 16

    Non protein kilocalories-to-nitrogen ratio:

       99.6 * 4 = 398.4 protein calories

$$1800 - 398 = 1401 \text{ non protein calories}$$

$$\frac{\text{non protein kcalories}}{\text{nitrogen}} = \frac{1401}{16} = 87.6$$

The total kilocalories-to-nitrogen ratio:

$$\frac{\text{total kcalories}}{\text{nitrogen}} = \frac{1800}{16} = 112.5$$

E. Diet order: Twocal HN 80 ML, 1/2-strength every hour

Total ml in 24 hours: 80 ml * 24 hours = 1920 ml

Total calories: 1920 ml * 2 calories per ml divided by .5 = 1920

Protein (in grams): 83.7 * 1920 divided by 1000 divided by 2 = 80.35

USRDA vitamins and minerals (percentage): 100 * 1920 divided by 1900 = 101.05%

Water from formula (in cc): 1920 cal * 57.7 = 1108 ml

Grams of nitrogen: 80.35 divided by 6.25 = 12.9

Non protein kilocalories-to-nitrogen ratio:

80.35 * 4 = 321.4 protein calories

1920 − 321 = 1599 non protein calories

$$\frac{\text{non protein kcalories}}{\text{nitrogen}} = \frac{1599}{12.9} = 124$$

The total kilocalories-to-nitrogen ratio:

$$\frac{\text{total kcalories}}{\text{nitrogen}} = \frac{1920}{12.9} = 149$$

F. Diet order: Vivonex Ten 100 ml, 3/4-strength every hour

Total ml in 24 hours: 100 ml * 24 = 2400 ml

Total calories: 2400 ml * 1 calorie * .75 = 1800

Protein (in grams): 38.2 * 1800 Cal divided by 1000 = 68.76

USRDA vitamins and minerals (percentage): 100 * (.75 * 2400) divided by 2000 = 90%

Water from formula (in cc): (.75 * 2400 cal) * .85 = 1530

Grams of nitrogen: 68.76 divided by 6.25 = 11

Non protein kilocalories-to-nitrogen ratio:

68.76 * 4 = 275 protein calories

1800 − 275 = 1525 non protein calories

$$\frac{\text{non protein kcalories}}{\text{nitrogen}} = \frac{1525}{11} = 138.6$$

The total kilocalories-to-nitrogen ratio:

$$\frac{\text{total kcalories}}{\text{nitrogen}} = \frac{1800}{11} = 163.6$$

Chowdhury—Electrical Engineering

Problem 95.   **A.** You have three measurements, therefore the mean is:

$$\bar{x} = \frac{99.6 + 101.5 + 100.1}{3} = 100.4$$

B. Sample variance:

$$\text{var } x = \frac{(99.6 - 100.4)^2 + (101.5 - 100.4)^2 + (100.1 - 100.4)^2}{3 - 1} = 0.97$$

Sample standard deviation $= \sqrt{0.97} = 0.9849$.

C. One percent of the mean $= 0.01 * 100.4 = 1.004$.

The sample standard deviation (0.9849) is less than one percent of the mean. Therefore, the specification for the resistance can be written as:

$$\text{resistance} = 100.4 \pm 3 * 0.9849 = 100.4 \pm 2.95 \text{ Ohms.}$$

D. Let the (yet unknown) best estimate be u. Then, the errors at the measured values are $(x_1 - u), (x_2 - u), \ldots, (x_n - u)$ etc. Total squared error is:

$$\text{err} = f(u) = (x_1 - u)^2 + (x_2 - u)^2 + \cdots + (x_n - u)^2$$

Non-calculus solution (recall the hint in the problem statement): First show by completing the square that $f(x) = ax^2 + bx + c$, a greater than zero, has its minimum value at $x = -b/2a$. Then consider

$$\text{err} = f(u) = (x_1 - u)^2 + (x_2 - u)^2 + (x_3 - u)^2$$

The non-calculus solution is:

$$f(x) = ax^2 + bx + c, \qquad a > 0$$

$$= a\left(x^2 + \frac{b}{a}x + \frac{c}{a}\right)$$

$$= a\left(x^2 + \frac{b}{a}x + \frac{b^2}{4a^2}\right) + \left(c - \frac{b^2}{4a}\right)$$

$$= a\left(x + \frac{b}{2a}\right)^2 + \left(c - \frac{b^2}{4a}\right)$$

which is a minimum when $x = -\frac{b}{2a}$.

$$\text{err} = f(u) = (x_1 - u)^2 + (x_2 - u)^2 + (x_3 - u)^2$$

$$= 3u^2 - u(2x_1 + 2x_2 + 2x_3) + (x_1^2 + x_2^2 + x_3^2).$$

Because $f(u)$ is a quadratic function of u, its minimum occurs at:

$$\frac{-[-(2x_1 + 2x_2 + 2x_3)]}{2 * 3}$$

which equals

$$\frac{x_1 + x_2 + x_3}{3} = \bar{x}$$

Note: This idea is easily generalized to n observations.

The calculus solution is: To find the minimum, take the first derivative of the error function and set that to zero:

$$f'(u) = -2(x_1 - u) - 2(x_2 - u) - \cdots - 2(x_n - u) = 0$$

or, $\quad x_1 + x_2 + \cdots + x_n - (u + u + \cdots + u) = 0$

i.e., $\quad x_1 + x_2 + \cdots + x_n - n * u = 0$

so that $\quad u = \dfrac{x_1 + x_2 + \cdots + x_n}{n} = \bar{x}$

The second derivative of $f(u)$ is positive ($f''(u) = 2n$), so the minimum of the function occurs at this value of u. Surprise! The best estimate according to this criterion turns out to be the same as the average value, which you have already calculated.

Problem 96. **i.** Failure probability of Tl is:

$$\Pr(\text{Tl fails}) = 1 - \Pr(\text{Tl works}) = 1 - 0.89 = 0.11.$$

Therefore:

$$\Pr(\text{Tl fails, T2 works}) = 0.11 * 0.96 = 0.1056.$$

ii. Failure probability of T2 is $1 - 0.96 = 0.04$. Therefore:

$$\Pr(\text{T2 fails, Tl works}) = 0.04 * 0.89 = 0.0356.$$

iii. $\Pr(\text{Tl fails, T2 fails}) = 0.11 * 0.04 = 0.0044.$

iv. "At least one of them works" can happen in the following ways: Both Tl and T2 work, or Tl works but T2 does not, or T2 works but Tl does not. So:

$$\Pr(\text{Both Tl and T2 work}) = 0.89 * 0.96 = 0.8544.$$

You have already found the other two probabilities. Adding them, you get:

$$\Pr(\text{At least one works}) = 0.8544 + 0.0356 + 0.1056 = 0.9956.$$

However, an easier way to calculate this is to realize that:

$$\Pr(\text{At least one works}) = 1 - \Pr(\text{both Tl and T2 fail}) = 1 - 0.0044 = 0.9956.$$

Problem 97. To find a maximum or minimum, take the derivative of $f(x)$ and set it to zero.

$$f(x) = -x^2 + 18x + 12$$

$$f'(x) = -2x + 18 = 0$$

solving for x, get $x = 9$. To decide whether $x = 9$ is a maximum, take the second derivative and check its sign. If it is negative, it is a maximum. In this case:

$$f''(x) = -2$$

Therefore, at $x = 9$ (9:00 a.m.) the maximum efficiency occurs. The value of this efficiency is given by:

$$f(x = 9) = 93\%$$

Alternatively, you can graph the parabola $-x^2 + 18x + 12$, noting that its vertex occurs at $x = -18/(-2) = 9$, and that the vertex is the highest point on the parabola.

Dinkey—Chemical Engineering

Problem 98. **A.** The mean equals $\bar{x}$.

$$\bar{x} = (2.63 + 2.72 + 2.47 + 2.55)/4$$

$$= 10.37/4$$

$$= 2.592 = 2.59 \text{ weight-\% phosphorus.}$$

B. The individual deviations from the mean are:

Run Number	Value	Deviations
1	2.63	−.04
2	2.72	−.13
3	2.47	.12
4	2.55	.04

C. Given the formula

$$s = \sqrt{\frac{(x_1 - \bar{x})^2 + (x_2 - \bar{x})^2 + (x_3 - \bar{x})^2 + (x_4 - \bar{x})^2}{3}}$$

Use a table like the following to help keep track of your calculations:

Run Number	Value	Deviations	(Deviations)²
1	2.63	.04	.0016
2	2.72	.13	.0169
3	2.47	−.12	.0144
4	2.55	−.04	.0016

Therefore, the standard deviation, s, of these runs is:

$$s = \sqrt{\frac{.0016 + .0169 + .0144 + .0016}{3}}$$

$$= \sqrt{\frac{.0345}{3}} = \sqrt{.0155} = .1072 \text{ wt.-\% P.}$$

Problem 99. The standard deviation, s, of these runs is calculated as follows:

Run Number	Value	Deviations	(Deviations)^2
1	1.09	-1.5	2.25
2	3.06	$-.47$	.221
3	4.10	.47	.221
4	2.12	1.51	2.28

$$s = \sqrt{\frac{2.25 + .221 + .221 + 2.28}{3}}$$

$$= \sqrt{\frac{4.97}{3}} = \sqrt{1.657} = 1.287 \text{ wt.-\% P.}$$

The standard deviation of this set of points is much greater than the standard deviation for the first set, because the spread of the data is greater. The smaller standard deviation indicates a more reproducible process.

LoVerso—Software Engineering

Problem 100. **a.** The stripe size is simply $7 * 16 = 112$ bytes.

b. For this problem, use a modulo function to get the remainder. It does not matter how many complete stripes are written before the last one, so take the modulo to figure out how far into the last stripe to go: $4321 \bmod 112 = 65$. So, the final stripe that is written contains 65 bytes. Divide that amount by the amount of data per drive to find out how many drives are completely written to: $65/16 = 4.06$. That means four drives are written to completely, and a little bit (one byte, actually) is written to the fifth drive in the device. Since the drives start numbering at 0, we end up on Disk 4.

c. Apply the same techniques as the second part of this problem: $8000 \bmod 112 = 48$. Start 48 bytes into the stripe: $48/16 = 3$. Skip three disks and begin the I/O on the fourth disk (Disk 3).

To determine the ending location, add the offset to the amount written and compute where it ends:

$$8000 + 1500 = 9500$$

$$9500 \bmod 112 = 92 \text{ bytes in the last stripe}$$

$$92/16 = 5.75$$

So, writing ends on the sixth disk, or Disk 5. For the second part of the problem, simply adjust the values from above. Start the write operation 48 bytes into a stripe. Therefore, write $112 - 48$ or 64 bytes to the first stripe before beginning back at Disk 0. We also know that we end up 92 bytes into the last stripe at the end of the operation. Subtract those two partial pieces for the total size: $1500 - (64 + 92) = 1344$ bytes. This means that 1344 bytes worth of complete stripes are completed for this write operation. And yes, $1344 \bmod 112 = 0$, so it does evenly divide. To compute how many complete stripes to write, divide: $1344/112 = 12$ complete stripes.

Problem 101. First, draw the layout of the data and represent the data in its binary form (for example, base 2). The first nine numbers are 1, 1, 2, 3, 5, 8, 13, 21, 34. Writing one number per disk creates the following layout:

Disk 0	Disk 1	Disk 2
1	1	2
3	5	8
13	21	34

To do an XOR on those numbers, write them in binary form:

Disk 0	Disk 1	Disk 2
00000001	00000001	00000010
00000011	00000101	00001000
00001101	00010101	00100010

Compute parity by iteratively XOR-ing across each row:

$$((00000001 \text{ XOR } 00000001) \text{ XOR } 00000010) = 00000010$$
$$((00000011 \text{ XOR } 00000101) \text{ XOR } 00001000) = 00001110$$
$$((00001101 \text{ XOR } 00010101) \text{ XOR } 00100010) = 00111010$$

Now convert those numbers back to decimal from binary:

$$00000010 = 2$$
$$00001110 = 15$$
$$00111010 = 58$$

Problem 102. Do the same computation for this problem, but use the parity information to compute what is on Drive 1. First, convert the numbers into binary form:

Disk 0	Disk 2	Parity
00010101	00000000	00000010
00100010	00010000	01000101
00110111	10010000	11111110

XOR across each row:

$$((00010101 \text{ XOR } 00000000) \text{ XOR } 00000010) = 00010111$$
$$((00100010 \text{ XOR } 00010000) \text{ XOR } 01000101) = 01110111$$
$$((00110111 \text{ XOR } 10010000) \text{ XOR } 11111110) = 01011001$$

Now convert those numbers to decimal:

$$00010111 = 23$$
$$01110111 = 119$$
$$01011001 = 89$$

Problem 103. These numbers are represented as:

$$
\begin{aligned}
abcabc = \quad & 100{,}000a + 10{,}000b + 1000c \\
+ \quad & \underline{\phantom{00{,}00}100a + 10b + 1c} \\
& 100{,}100a + 10{,}010b + 1001c = 1{,}001(100a + 10b + c)
\end{aligned}
$$

All those six-digit numbers are multiples of 1001. Since 1001 is evenly divisible by seven, all multiples will also be divisible by seven.

Problem 104. First calculate what the total EPS must be for this year to meet the 15 percent increase:

$$\$1.50 * 1.15 = 1.72.$$

Sum up the first three quarters and subtract from 1.72 to get the EPS required in the 4th quarter:

$$1.72 - (.33 + .40 + .48) = 1.72 - 1.21 = .51.$$

Multiply the EPS by the number of outstanding shares to get the overall earnings of the company:

Previous year: $1.50 * 48$ million $= 72$ million
This year: $1.72 * 48$ million $= 82.56$ million

Problem 105. First, compute how many shares are now outstanding:

$$48,000,000 - (48,000,000 * .08) = 48,000,000 - 3,840,000 = 44,160,000.$$

Then, compute earnings per share:

$$82,560,000/44,160,000 = \$1.86 \text{ EPS}.$$

Now, calculate the increase over last year:

$$1.86/1.24 = 24\%$$

(I would be thrilled with a return of 24 percent on my money!)

Thatcher—Immunology and Microbiology

Problem 106. 1 Tbs.   8 oz./16 = 1/2 oz. = 1 Tbs.

Problem 107. Two gallons, or 32 eight-ounce glasses.

$$1 \text{ lb.} = 16\text{oz.};$$

$$16\text{oz.} \times 16 \text{ (dilution factor)} = 256 \text{ oz.}$$

$$1 \text{ gal.} = 128 \text{ oz.};$$

$$256 \text{ oz.}/128 \text{ oz.} = 2 \text{ gal.}$$

or

$$1 \text{ lb.} = 16 \text{ oz.} = 32 \text{ Tbs.}$$

$$1 \text{ Tbs.}/8\text{-oz. glass} \times 32 \text{ Tbs.} = 32 \text{ eight-oz. glasses}$$

Problem 108. 45.5 ml 11 M hydrochloric acid; 954.5 ml water

Using $C_1V_1 = C_2V_2$, first assign values:

For acid: $\quad C_1 = 11M$; $V_1 = $ unknown

$\qquad\qquad C_2 = 0.5M$; $V_2 = 1000$ ml

$\qquad\qquad V_1 = C_2V_2/C_1$

$\qquad\qquad V_1 = 0.5$ M $\times$ 1000 ml/11 M

$\qquad\qquad V_1 = 45.5$ ml

For water: $\quad$ 1000 ml total volume $- 45.5$ ml acid $= 954.5$ ml water.

Note that the total final volume must equal 1000 ml. Therefore the volume of concentrated acid is subtracted from 1000 ml to find the amount of pure water needed.

Problem 109. Final dilution for tube #4 $= 1/10,000$ or 10^{-4}.

Note: Four tubes each have a ten-fold dilution. The final concentration is equal to the dilution factor, ten, to the negative power of the number of tubes, -4. This works for any dilution factor and any number of tubes, and is very useful in practice.

$$\frac{1 \text{ ml sample}}{10 \text{ ml in tube \#1}} = 0.1 \text{ ml sample/ml}$$

$$\frac{1 \text{ ml of \#1}}{10 \text{ ml in tube \#2}} = \frac{0.1 \text{ ml sample}}{10 \text{ ml}} = 0.01 \text{ ml sample/ml}$$

$$\frac{1 \text{ ml of \#2}}{10 \text{ ml in tube \#3}} = \frac{0.01 \text{ ml sample}}{10 \text{ ml}} = 0.001 \text{ ml sample/ml}$$

$$\frac{1 \text{ ml of \#3}}{10 \text{ ml in tube \#4}} = \frac{0.001 \text{ ml sample}}{10 \text{ ml}} = 0.0001 \text{ ml sample/ml}.$$

The amount of initial sample in tube #4 $= 0.001$ ml, a very small amount indeed. To get the final $1/10,000$ dilution, $1/1000$ of the initial one ml of sample would have to be added to the last tube of nine ml water. That is too small to accurately measure using standard lab pipettes. As you can see, dilution series allows you to achieve accuracy in measuring small amounts.

Problem 110. 5200/ml, or 5.2×10^3

260 colonies/ 5 ml dilution sample $= 52$ colonies/ml dilution sample.

52 colonies $\times$ 100 (dilution factor) $= 5200$/ml.

[There are too few colonies on plates #3 and #4 to give an accurate estimate of bacteria in the sample.]

The sample contained about 5200 bacterial cells per milliliter. Whether this amount of bacteria is significant depends on the type of bacteria found and the source of the water tested.

Problem 111. The final dilution in tube #4 $= 1/16$.

Add equal amounts of sample and diluent. For example, if each tube contained five ml diluent, add five ml of sample to the first tube. Then transfer five ml of mixture to the next, and so on until finished. Two-fold dilution in four tubes can be expressed as 2^{-4}, equal to a

1/16 final dilution.

$$\text{Tube \#1} = \frac{5 \text{ ml sample}}{10 \text{ ml total}} = 1/2 \text{ dilution of sample}$$

$$\text{Tube \#2} = \frac{5 \text{ ml tube \#1}}{10 \text{ ml total}} = 1/4 \text{ dilution of sample}$$

$$\text{Tube \#3} = \frac{5 \text{ ml tube \#2}}{10 \text{ ml total}} = 1/8 \text{ dilution of sample}$$

$$\text{Tube \#4} = \frac{5 \text{ ml tube \#3}}{10 \text{ ml total}} = 1/16 \text{ dilution of sample.}$$

The mathematics of dilutions are not difficult, but many college students find the concepts hard to grasp. Some complain it is because they are not good at math or chemistry. However, a little practice in analysis and thinking about what is happening in those tubes is all that is needed. By learning how to set up and solve dilution problems, you will save yourself many hours of frustration in college, as well as at work and at home. Have some more chocolate milk—you earned it!

Pollitt—Mechanical Engineering

Problem 112. It is given that $RPQ = 45°$, and if a perpendicular is dropped from point Q to the centerline of the duct on the ground, then $QRP = 90°$. A geometry theorem states that the sum of all the angles in a triangle is $180°$. Hence, $PQR = 45°$. This forms an isosceles triangle because the two sides, QR and RP, are both equal.

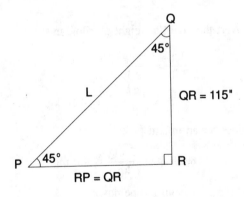

FIGURE 84

Using the Pythagorean Theorem: $\overline{QR}^2 + \overline{RP}^2 = L^2$

$$L = R(\overline{QR}^2 + \overline{RP}^2) = 162.6 \text{ inches (13.55 feet).}$$

Townsend-Beteet—HMO Pharmacy Practice and Management

Problem 113. Calculate the number of tablets needed for a one-month supply:

$$\frac{3 \text{ mg}}{\text{dose}} * \frac{4 \text{ doses}}{\text{day}} * 30 \text{ days} = 360 \text{ mg, or 36 tablets.}$$

Next, calculate the final volume of cherry syrup. The desired concentration is three mg per teaspoon (or per five ml). Set up a ratio:

$$\frac{5 \text{ ml}}{3 \text{ mg}} = \frac{x \text{ ml}}{360 \text{ mg}}.$$

Solve for x:

$$3 \text{ mg} * x \text{ ml} = 5 \text{ ml} * 360 \text{ mg}$$
$$x \text{ ml} = \frac{1800 \text{ ml mg}}{3 \text{ mg}}$$
$$x = 600 \text{ ml.}$$

The tablets are crushed and the cherry syrup is added to make the final volume.

Problem 114. Calculate how many tablets are needed for 15 days:

$$\frac{1 \text{ mg}}{\text{dose}} * \frac{4 \text{ doses}}{\text{day}} * 15 \text{ days} = 60 \text{ mg, or six tablets.}$$

Calculate the final volume of cherry syrup. The desired concentration is one mg per 2.5 ml. Set up a ratio:

$$\frac{2.5 \text{ ml}}{1 \text{ mg}} = \frac{x \text{ ml}}{60 \text{ mg}}.$$

Solve for $x = 150$ ml.

Problem 115. First convert the infant's weight to kilograms:

$$\frac{1 \text{ kg}}{2.2 \text{ lb}} = \frac{x \text{ kg}}{18 \text{ lb}}$$
$$x = \frac{18}{2.2} \text{ kg} = 8.18 \text{ kg.}$$

Then calculate the safe dose for an infant:

$$8.18 \text{ kg} * \frac{.5 \text{ mg}}{\text{kg}} = 4.0 \text{ mg.}$$

The infant could take 4.0 mg of Trispin in one day.

Problem 116. $1.9 \text{ m}^2 * 40 \text{ mg/m}^2 = 76$ mg, or 7.5 tablets.

Problem 117. Find the amount of Compound A to be present in the final solution.
Percentage desired = 20 percent, i.e.,
20 percent of 120 ml = $0.20 * 120$ ml = 24 ml.

How much of the stock solution will provide 24 grams?

$$50\% \text{ of } y = 24 \text{ ml}$$

$$.5y = 24 \text{ ml}$$

$$y = 48 \text{ ml.}$$

Distilled water is determined by subtraction:

$$\begin{array}{l} 120\text{ml Total volume} \\ \underline{- 48\text{ml Amount of solute}} \\ 72\text{ml Amount of water to add} \end{array}$$

So, adding 48 ml of the 50 percent solution of Compound A to 72 ml of distilled water will produce 120 ml of 20 percent solution of Compound A, as prescribed.

Problem 118. Using the same steps as in problem 5, figure:

$$10\% \text{ of } 120 \text{ ml} = 0.10 * 120\text{ml} = 12 \text{ ml.}$$

Solve for how much stock solution to provide:

$$50\% \text{ of } y = 12 \text{ ml}$$

$$.5 \, y = 12 \text{ ml}$$

$$y = 24 \text{ ml, the amount of solute.}$$

$$\begin{array}{l} 120\text{ml Total volume} \\ \underline{- 24\text{ml Amount of solute}} \\ 96\text{ml Amount of water to add} \end{array}$$

So, adding 24 ml of the 50 percent solution of Compound A to 96 ml of distilled water will produce 120 ml of 10 percent solution of Compound A, as prescribed.

Problem 119. $60 * .05 = 3.0$ ml of LCD solution.
$60 * .05 = 3.0$ mg of aspirin powder.
Tiamcinolone solution = approx. 54 ml.

Kivlin—Ophthalmology

Problem 120. a. Thirteen prism diopters at 22^+ degrees from horizontal. The power is calculated by the Pythagorean theorem. The angle is found from the tangent of the angle, because the two legs of the triangle are known.

   b. Six and one-half prism diopters at 22^+ degrees from horizontal in each eye. The optician who grinds the glasses will need to know which lens has the prism oriented with its base up (left), and which has the prism base down (right).

Problem 121. a. Accommodation = $100/28.5$ cm = 3.5 diopters.

b. Accommodation = 100/57 cm = 1.75 (from the table, this is age 55 years).

c. 100/33 cm = three diopters needed for that distance. Subtract one diopter for the accommodation that still remains, so the result is two diopters glasses power.

Problem 122. **a.** 1000 mg/100 ml = 10 mg/ml. 10 mg/20 drops/ml = 1/2 mg per drop. Therefore, 2 drops = 1 mg total.

b. 1 mg × 10% = 0.1 mg.

c. 0.07 mg. This is less than what the baby would receive from the drops. So the drops can deliver a large dose of Atropine, a very powerful drug.

Lum—Electrical Engineering

Problem 123. See Figure 85.

First, the value of x must be calculated. This problem can be solved using the Pythagorean Theorem:

$$a^2 + b^2 = c^2.$$

Using the Pythagorean Theorem and algebra to solve for x:

$$a^2 + b^2 = c^2$$
$$R_{\text{max}}^2 + L_h^2 = (R_{\text{max}} + x)^2$$
$$\sqrt{R_{\text{max}}^2 + L_h^2} = R_{\text{max}} + x$$

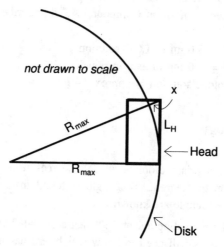

Where x = distance scanner moves per disk revolution

FIGURE 85

Distance scanner moves per disk revolution

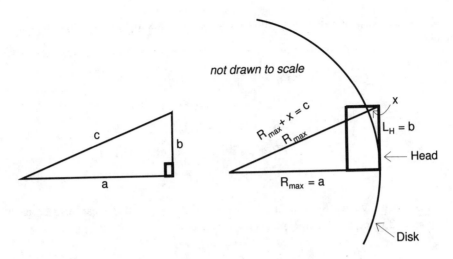

FIGURE 86
Pythagorean Theorem applied to Figure 85

$$x = \sqrt{R_{\max}^2 + L_h^2} - R_{\max}$$

$$x = \sqrt{(6.75 \text{ in})^2 + (0.16 \text{ in})^2} - 6.75 \text{ in}$$

$$x = 0.0019 \text{in}.$$

The equation for the scan speed can be derived by realizing that the head needs to move some number of in/min (inches per minute).

Since the head needs to move x in/rev, and the disk spins at v rev/min,

$$(\text{in/rev}) * (\text{rev/min}) = (\text{in/min}).$$

Thus,

$$\text{Scan speed} = (x \text{ in/rev})(v \text{ rev/min})$$

$$= (0.0019 \text{ in/rev})(3600 \text{ rev/min})$$

$$= 6.84 \text{ in/min}.$$

Problem 124. To solve this problem, apply Kirchoff's Voltage Law (KVL) and Ohm's Law. Using Ohm's law:

$$V_1 = IR_1 \qquad \text{(Equation 1)}$$

$$V_{\text{out}} = IR_2 \qquad \text{(Equation 2)}.$$

Using KVL,

$$-V_{\text{in}} + V_1 + V_{\text{out}} = 0 \qquad \text{(Equation 3)}.$$

Substituting Equation 1 for V_1,

$$-V_{\text{in}} + IR_1 + V_{\text{out}} = 0 \qquad \text{(Equation 4)}.$$

Since V_{in}, I, and V_{out} are given, Equation 4 can be used to solve for R_1 after some algebraic manipulation.

$$-V_{in} + IR_1 + V_{out} = 0$$
$$IR_1 = V_{in} - V_{out}$$
$$R_1 = (V_{in} - V_{out})/I$$
$$R_1 = (15 \text{ volts} - 4 \text{ volts})/0.01 \text{ amps}$$
$$R_1 = 1100 \text{ W.}$$

Similarly, since the values for V_{out} and I are given, the value of R_2 can also be calculated from Equation 2 using a little algebra.

$$V_{out} = IR_2$$
$$R_2 = V_{out}/I$$
$$R_2 = 4 \text{ volts}/0.01 \text{amps}$$
$$R_2 = 400 \text{ W.}$$

Problem 125. **a.** $8 = 1000_2$ $\quad 3 = 11_2$ $\quad 12 = 1100_2$ $\quad 5 = 101_2$

b. $8 = 10_8$ $\quad 3 = 3_8$ $\quad 12 = 14_8$ $\quad 25 = 31_8$

c. $8 = 8_{16}$ $\quad 3 = 3_{16}$ $\quad 12 = C_{16}$ $\quad 20 = 14_{16}$

d. $1101_2 = 13$ $\quad 10000_2 = 16$ $\quad 17_8 = 15$ $\quad 20_8 = 16$ $\quad E_{16} = 14$ $\quad 2D_{16} = 45$

e.

1010	1010	101	1111	1011
+ 100	+ 10	+ 1101	+ 11	+ 110
1110	1100	10010	10010	10001

MacConnell—Fish Pathology

Problem 126. $(6.0 * 60 * 5.0 * 0.0038)/1.00 = 6.84$ grams of chemical B

Problem 127. $(300 * 60 * 2.0 * 0.0038)/0.80 = 171$ ml of chemical C

Problem 128. **a.** Using the formulas given:

$$\% \text{ Saturation} = \frac{BP + DP}{BP} * 100$$
$$= \frac{587 + 38}{587} * 100$$
$$= 106.5\% \text{ saturation.}$$

$$\% \text{ Nitrogen } = \frac{(BP + DP) - \left[\frac{DO}{BC}(CF1)\right] - VP}{(BP - VP)CF2} * 100$$

$$= \frac{(587 + 38) - \left[\frac{7.2}{0.035}(0.532)\right] - 11.98}{(587 - 11.98)0.790} * 100$$

$$= \frac{625 - 109.4 - 11.98}{454.3} * 100$$

$$= \frac{503.6}{454.3} * 100$$

$$= 110.8\% \text{ N.}$$

b. Using the formula for %N again:

$$\% \text{ Nitrogen } = \frac{(587 + 35) - \left[\frac{10}{0.038}(0.532)\right] - 9.2}{(587 - 9.2)0.790} * 100$$

$$= \frac{622 - 139.9 - 9.2}{456.5}$$

$$= \frac{472.9}{456.5} * 100$$

$$= 103.6\%$$

Any saturation over 100 percent can pose a threat to fish health, so the nitrogen could be a problem, but many species of fish can tolerate this level.

Swetman—Computer Science and Computer Graphics

Problem 129.

$$M = 77_{10} = 4D_{16} = 0100\ 1101$$

$$a = 97_{10} = 61_{16} = 0110\ 0001$$

$$t = 116_{10} = 74_{16} = 0111\ 0100$$

$$h = 104_{10} = 68_{16} = 0110\ 1000$$

Concatenated together, the computer would store Math as:

$$0100\ 1101\ 0110\ 0001\ 0111\ 0100\ 0110\ 1000$$

Notice how much easier it is to calculate using hex:

$$M = 77_{10} = (4_{16} * 16^1) + (D_{16} * 16^0) = 4D = 0100\ 1101$$

and since $4_{16} = 0100$ and $D_{16} = 1101$, $4D = 0100\ 1101$.

By contrast:

$$M = 77_{10} = (1 * 2^6) + (1 * 2^3) + (1 * 2^2) + (1 * 2^1)$$
$$= (0 * 2^7) + (1 * 2^6) + (0 * 2^5) + (0 * 2^4) + (1 * 2^3)$$
$$+ (1 * 2^2) + (1 * 2^1) + (0 * 2^0)$$
$$= 0100\ 1110.$$

Problem 130.   The hex values 62 69 6E 61 72 79 represent the word "binary" in ASCII. The binary representation of the word "binary" is:

$$0110\ 0010\ 0110\ 1001\ 0110\ 1110\ 0110\ 0001\ 0111\ 0010\ 0111\ 1001$$

Here are the values in binary, hex, decimal, and ASCII letters:

$$0110\ 0010 = 62_{16} =\ 98_{10} = b$$
$$0110\ 1001 = 69_{16} = 105_{10} = i$$
$$0110\ 1110 = 6E_{16} = 110_{10} = n$$
$$0110\ 0001 = 61_{16} =\ 97_{10} = a$$
$$0111\ 0010 = 72_{16} = 114_{10} = r$$
$$0111\ 1001 = 79_{16} = 121_{10} = y$$

This problem stated the ASCII values for upper-case and lower-case letters to be 65 to 90 and 97 to 122 (decimal) respectively, but actually these values are more often given in hexadecimal. It is almost as easy to count and do simple addition in hex as it is in decimal once you get used to it. So "a" $= 61_{16}$ must mean that "i" $= 69_{16}$ and "j" $= 6A_{16}$, et cetera.

Problem 131.   Here is a short sample name: Jill Mann

$$J = 74_{10} = 4A_{16} = 0100\ 1010$$
$$i = 105_{10} = 69_{16} = 0110\ 1001$$
$$l = 108_{10} = 6C_{16} = 0110\ 1100$$
$$l = 108_{10} = 6C_{16} = 0110\ 1100$$
$$= 32_{10} = 20_{16} = 0010\ 0000$$
$$M = 77_{10} = 4D_{16} = 0100\ 1101$$
$$a = 97_{10} = 61_{16} = 0110\ 0001$$
$$n = 110_{10} = 6E_{16} = 0110\ 1110$$
$$n = 110_{10} = 6E_{16} = 0110\ 1110$$

Or strung together:

$$0100\ 1010\ 0110\ 1001\ 0110\ 1100\ 0110\ 1100\ 0010\ 0000$$
$$0100\ 1101\ 0110\ 0001\ 0110\ 1110\ 0110\ 1110$$

Problem 132.   black $= (0, 0, 0)$
                           white $= (1, 1, 1)$

Problem 133.   red $= (1, 0, 0)$
                           green $= (0, 1, 0)$
                           blue $= (0, 0, 1)$

Problem 134.   cyan $= (0, 1, 1)$
                           magenta $= (1, 0, 1)$
                           yellow $= (1, 1, 0)$

Problem 135. (x, y, z) where $x = y = z$ with values between zero and one. For instance, a
very light grey would be $(.001, .001, .001)$. A very dark grey, almost black, would be $(.9, .9, .9)$.

Moore—Mathematics and Computing

Problem 136.   The formula for the number of cells C at any time t is $C = 2^{2t}$.

Problem 137.   The formula for the number of cells at any time t is $C = 100 * 2^{2t}$. Using the
notation $C_0 = 100$, this becomes $C = C_0 2^{2t}$.

Problem 138.   Since the general formula is $C(t) = C_0 2^{kt}$ and the formula for the example in
Problem 137 is $C_0 2^{2t}$, $C_0 2^{kt} = C_0 2^{2t}$ means that $k = 2$, so the doubling time is $1/k = 1/2$.

Problem 139.   Use the formula to solve for $1/k$, the doubling time:

$$C = C_0 2^{kt}$$
$$160{,}000 = 10{,}000 * 2^{k3}$$
$$16 = 2^{k3}.$$

Note that $16 = 2^4$, so $16 = 2^{k3} = 2^4$ means:

$$k3 = 4$$
$$1/k = 3/4 \text{ hour.}$$

Problem 140.   Use the formula to solve for $1/k$, the doubling time:

$$C = C_0 2^{kt}$$
$$150{,}000 = 10{,}000 * 2^{k3}$$
$$15 = 2^{k3}$$
$$\log_2(15) = k3.$$

Note: to calculate $\log_2(15)$, use:

$$\log_2(15)\log_{10}(2) = \log_{10}(15)$$

$$log_2(15) = \frac{1.176}{.301} = 3.9.$$

So to continue the solution:

$$3.9 = k3$$

$$1.3 = k$$

$$1/k = .77 \text{ hour.}$$

Stiglich—Electrical Engineering

The answers to the problems are broken down into steps.

Problem 141. Step a. Draw a "mirror image" of the test setup. The equivalent heights and distances can be labeled as h', a', H' and L', as shown. Notice that $h = h'$, $H = H'$, $a = a'$, and $L = L'$.

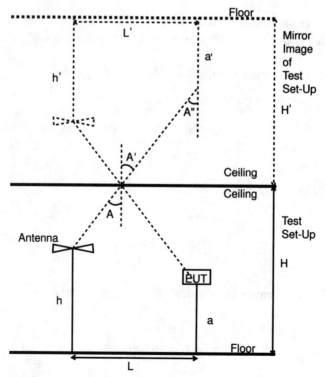

FIGURE 87
Mirror image of test setup

Step b. Using geometry, we see that $\angle A' = \angle A$, because opposite angles formed by two intersecting lines are equal. And $\angle A'' = \angle A$, because angles formed by a line intersecting two parallel lines are equal.

Step c. Next, identify the similar triangles that contain $\angle A$ and $\angle A''$. They are outlined in Figure 88. These triangles are similar triangles because their angles are identical. Label the sides of the large triangle containing A'' as x, y, and z.

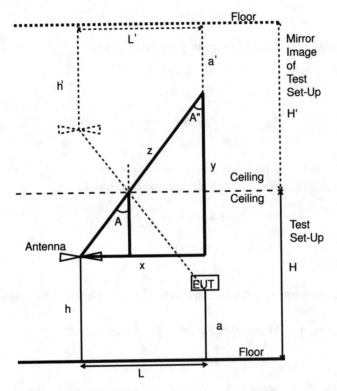

FIGURE 88
Similar triangles

Step d. Find expressions for sides x and y in terms of a, h, H and L. Notice that $x = L$ and $y = 2H - a' - h$. Since $a' = a$, $y = 2H - a - h$.

Step e. Find an expression for the tangent of Angle A''.

$$\tan A'' = \frac{\text{opposite}}{\text{adjacent}} = \frac{x}{y} = \frac{L}{2H - a - h}.$$

Step f. Since $\angle A'' = \angle A$, we can write

$$\tan A = \frac{L}{2H - a - h}.$$

Step g. To solve for angle A, take the arc tangent of both sides of the equation. Remember that $\arctan(\tan A) = A$. So

$$A = \arctan\left(\frac{L}{2H - h - a}\right).$$

Problem 142. **Step a.** The equation for the angle of incidence is

$$A = \arctan\left(\frac{L}{2H - h - a}\right).$$

Step b. Identify L, H, h, and a from the problem above:

$$L = 1m$$

$$H = 2m$$

$$h = 1m$$

$$a = 1m$$

Note that all distances must be in the same units, meters (m).

Step c. Plug the numbers into the formula and compute A.

$$A = \arctan\left(\frac{1}{2 * 2 - 1 - 1}\right)$$

$$= \arctan 1/2$$

$$= 26.56°.$$

So A is less than 45 degrees, and a ceiling height H of two meters is adequate.

Problem 143. **Step a.** The equation for the angle of incidence is:

$$A = \arctan\left(\frac{L}{2H - h - a}\right).$$

Step b. Identify L, H, h and a for each chamber, and fill in the table below. Hint: remember to convert feet to meters so all units stay the same.

$$12 \text{ feet} * 1 \text{ yard}/3 \text{ feet} * .9 \text{ meter}/1 \text{ yard} = 3.6 \text{ m}.$$

	L	H	h	a	(in meters)
Chamber #1	3	2	1	1	
Chamber #2	3	3.6	1	1	

Step c. Plug the numbers into the formula and compute A for each chamber.
Chamber #1:

$$A = \arctan\left(\frac{3}{2 * 2 - 1 - 1}\right)$$

$$= \arctan(3/2)$$

$$= \arctan(1.5)$$

$$= 56.3°.$$

Chamber #2:

$$A = \arctan\left(\frac{3}{2*3.6 - 1 - 1}\right)$$
$$= \arctan(3/5.2)$$
$$= \arctan(.58)$$
$$= 30.1°.$$

So Chamber #2 is acceptable for performing the test, because the angle of incidence A is less than 45 degrees.

Roman—Astronomy

Problem 144.   A. Determine the distance to the star from the sun: From the definition of magnitude, $m_2 - m_1 = 2.5 \log(l_1/l_2)$. Because the apparent brightness decreases as the inverse square of the distance:

$$l_1 r_1^2 = l_2 r_2^2$$

where

$$l_1 = \text{the luminosity of a star at the distance } r_1$$

and

$$l_2 = \text{the luminosity of the same star at a distance } r_2.$$

Hence:

$$m_2 - m_1 = 2.5 \log(r_2^2/r_1^2)$$
$$= 5 \log(r_2/r_1).$$

Since in this case, $m_2 - m_1 = 8.5 - (-1.5) = 10$

$$\log(r_2/r_1) = 2$$
$$r_2 = 100 r_1$$

or

$$r_2 = 100 * 3 * 10^{14}\text{km} = 3 * 10^{16}\text{km}$$

where

$$r_2 = \text{the distance of the star from the sun.}$$

B. Determine the position of the star with respect to the sun and the galactic center (see Figure 89).

By the law of cosines:

$$r_*^2 = (3 * 10^{16})^2 + (3 * 10^{17})^2 - 2 * 9 * 10^{33} * \cos(125°)$$
$$= 9 * 10^{32} + 9 * 10^{34} + 18 * 10^{33} * 0.57358$$

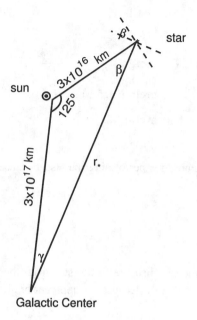

sun

3×10¹⁶ km

125°

β'

β

star

3×10¹⁷ km

r_*

γ

Galactic Center

FIGURE 89

The position of the star with respect to the sun and the galactic center

$$= 9.09 * 10^{34} + 1.032 * 10^{34}$$

$$= 10.122 * 10^{34}$$

or

$$r = 3.18 * 10^{17}$$
$$\sin g = \frac{3 * 10^{16} * \sin(125°)}{3.18 * 10^{17}} = 0.07728$$
$$g = 4°.4$$

and

$$\sin b = \frac{3 * 10^{17} * \sin(125°)}{3.18 * 10^{17}} = 0.7728$$
$$b = 50°.6.$$

The height of the star above the galactic plane is (see Figure 90):

$$h = 3 \times 10^{16} * \sin(10°) = 5.21 \times 10^{15} \text{km}.$$

Since the distance of the star from the galactic center is 3.18×10^{17}km, the inclination of the orbit is determined by:

$$\tan i = (0.521 \times 10^{16})/(3.18 \times 10^{17})$$

$$= 0.0164$$

$$i = 1°.$$

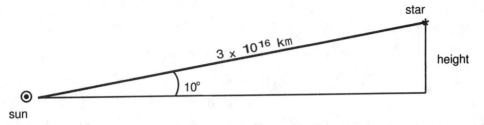

FIGURE 90
Geometry for height of the star above the galactic plane

The distance of the star from the center of the galaxy, projected on the galactic plane is: $3.18 \times 10^{17} \times \cos(1°) = 3.18 \times 10^{17}$km.

C. Determine the tangential velocity of the star (See Figure 58, p. 150). The proper motion of the star is $0''.02/\text{year} = 0.02/206265 = 9.74 * 10^{-8}$rad/year (206265 is the number of seconds in a radian). At a distance of $3 * 10^{16}$km, this corresponds to:

$$9.70 * 10^{-8} * 3 * 10^{16} = 2.909 * 10^9 \text{ km/year.}$$

The number of seconds in a year are: $365.25 * 24 * 60 * 60 = 3.156 * 10^7$. Hence, $2.909 * 10^9$ km/year $= (2.909 * 10^9)/(3.156 * 10^7) = 9.22 * 10$ km/sec.

D. Resolve the tangential velocity into components perpendicular to the galactic plane and, in the galactic plane, perpendicular and parallel to the galactic center.

By the law of cosines:

$$\cos(a) = \cos(110°)\cos(80°) + \sin(110°)\sin(80°) * \cos(175°)$$

$$= -.342 * .174 - .940 * .985 * .996 = -0.982$$

$$a = 169°.$$

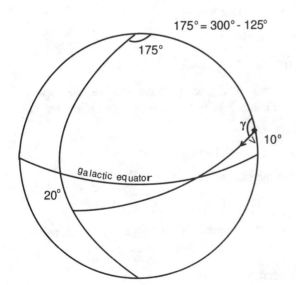

FIGURE 91
Geometry for resolving proper motion

By the law of sines:

$$sin(\gamma)/sin(110°) = sin(175°)/sin(169°);$$

$$sin(\gamma) = 0.429;$$

$$\gamma = 154°.6.$$

Note. The quadrant in this step and the next is determined from Figure 91 and logic. The motion perpendicular to the galactic plane is:

$$9.22 * cos(180° - \gamma) * 10 = 9.22 * 10 * cos(25°.4) = -83.1 \text{ km/sec.}$$

The motion parallel to the galactic plane is:

$$92.2 * sin(180° - \gamma) = 92.2 * sin(25°.4) = 39.6 \text{ km/sec.}$$

Referring to Figure 89, resolve the tangential velocity in the galactic plane into components parallel and perpendicular to the direction to the galactic center.

Motion perpendicular to the galactic center:

$$39.6 cos\beta = 39.6 cos(50°.6) = -25.1 \text{ km/sec}$$

Tangential motion parallel to the direction of the galactic center:

$$= 39.6 sin(50°.6)$$

$$= 30.6 \text{ km/sec.}$$

A study of a globe (or your grapefruit) will show that the velocity is in the same direction as that of the sun, i.e., positive.

E. Resolve radial velocity into components perpendicular to the galactic plane and, in the plane, perpendicular and parallel to the direction to the center. Latitude of the star as seen from the sun is ten degrees. Therefore, the component of radial velocity perpendicular to the plane is:

$$= -100 sin(10°) = -17.4 \text{ km/sec.}$$

The component of the radial velocity in the plane is:

$$= -100 cos(10°) = -98.5 \text{ km/sec.}$$

To resolve the component in the plane, again refer to Figure 89. Remember that the negative radial velocity is directed toward the sun. The component in the plane, parallel to the galactic center, is:

$$= -98.5 cos(50°.9) = -62.1 \text{ km/sec.}$$

The component in the plane, perpendicular to the galactic center, is:

$$= -98.5 sin(50°.6) = -76.4 \text{ km/sec.}$$

F. Compute components of the star's velocity perpendicular to the galactic plane, parallel and perpendicular to the galactic center. The velocity perpendicular to the plane is:

$$= -17.4 - 83.1 = -101 \text{ (that is, directed south) km/sec.}$$

The velocity in the plane perpendicular to the center is:

$$V_\perp = 235 - 76.4 - 25.1 = 133.5 \text{ km/sec.}$$

Remember, it is necessary to include the velocity of the sun. The velocity in the plane parallel to the direction to the center is:

$$V = 0 + 30.6 - 62.1 = -31.5.$$

G. Compute the total velocity in the plane:

$$V^2 = 31.5^2 + 133.5^2 = 18814$$

$$V = 137.2 \text{ km/sec.}$$

H. Determine the characteristics of the orbit. For the sun:

$$V^2 = 235^2 = GM/a \quad \text{(since } r = a \text{ for a circular orbit).}$$

Hence, $GM = 3 * 10^{17} * 235^2 = 1.657 * 10^{22}$.

For the star:

$$1/a = 2/r - V^2/GM$$
$$= 2/(3.18 * 10^{17}) - 137.2^2/(1657 * 10^{22})$$
$$= 6.29 * 10^{-18} - 1.136 * 10^{-18}$$
$$a = 1/(5.15 * 10^{-18})$$
$$= 1.94 * 10^{17} \text{ km.}$$

Also, $(r_* * V_\perp)^2 = GM * a * (1 - e^2)$ or

$$1 - e^2 = (3.18 * 133.5 * 10^{17})^2/(1.657 * 10^{22} * 1.94 * 10^{17}) = 0.56.$$

Therefore, $e = 0.66$. The perigalactic distance is

$$a(1 - e) = 1.94 * 10^{17} * 0.34 = 6.60 * 10^{16}.$$

The apogalactic distance is

$$a(1 + e) = 1.94 * 10^{17} * 1.66 = 3.22 * 10^{17}.$$

Claudia Zaslavsky—Author

Problem 145. The 12 ways to make two opening moves in Tic-Tac-Toe are shown in Figure 92.

If you are unconvinced, try coming up with a different, 13th opening two moves. Turn your new example around and over and compare it to the above 12 opening moves. It will match one of them.

Problem 146. Play the game.

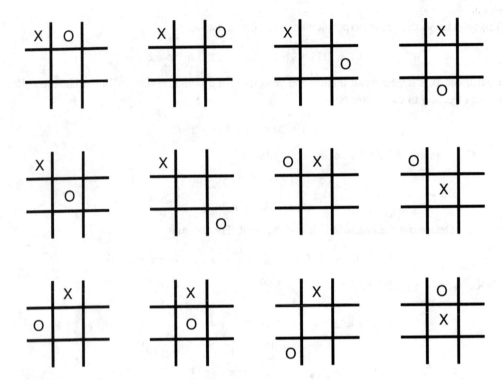

FIGURE 92
Twelve Opening Moves in Tic-Tac-Toe

Problem 147. Play the game.

Problem 148. I played the game both ways and decided that the nine-point version was the better game. However, when I posed the same question to groups of teachers and groups of students, they could not agree as to which version was better. Real-life problems are often like that, hard to resolve.

Problem 149. Discuss with your classmates whether playing with four counters for each player instead of three makes for a good game.

Taylor—Mathematics

Problems 150–154. Do it and find out! A part of an answer to the first problem is that, when you tape 3 or 4 or 5 triangles together along their edges so that they meet at one common vertex, they form a convex tent. Six triangles meeting at one point lie flat. Seven or more triangles ripple up and down. In problem 151, you find out that the sum of the angles at the vertex is the key: when it is less than 360 degrees (60 degrees per equilateral triangle corner), the surface is a convex tent; when the angle sum is exactly 360, the surface lies flat; and, when it is more than 360 degrees, the surface is "saddle-shaped" and not convex or flat. In problem 152 you should discover the regular convex polyhedra: the tetrahedron, octahedron, and icosahedron with triangles, the cube with squares, and the dodecahedron with regular pentagons. (For more information, see Coxeter's book *Regular Polytopes,* published by Dover.) Finally, a quantity called the Gauss curvature can be defined from the rate at which area grows with distance from a vertex. A book on differential geometry such as that by Frank Morgan is a source for further information on this topic, although it requires knowledge of multivariable calculus.